GRAVITATION
(According to 'Hypothesis on MATTER')

By:
Nainan K. Varghese

2010

TO
HIM

GRAVITATION
(According to 'Hypothesis on MATTER')
By: Nainan K. Varghese
Genre: Alternative Science - Theoretical Physics.
ISBN: 1450556264
EAN-13: 9781450556262

All rights reserved.

No part of this publication may be reproduced, stored in a retrieval system or transmitted in any form or by any means, electronic, mechanical, photocopying or otherwise, without the prior permission of the Author.

© Copyright: Nainan K. Varghese, 2010.

FOREWORD

Although we have many theories on gravitation, we are far from understanding its cause or mechanism of its actions. Contemporary theories deal with only a small part of gravitation – the (apparent) gravitational attraction between matter-bodies. Arguments of these theories are based on the worst assumption in science – 'the action at a distance' through empty space. Since 'action at a distance' is irrational, most conjectures about the cause and mechanism of actions of gravitation are also illogical.

'Hypothesis on MATTER' is an alternative concept that provides logical explanations to all physical phenomena of matter. It has no 'actions at a distance'. There is only one type of matter particles, postulated in this concept. These matter particles, in turn, produce everything else in this universe, including gravitation. Gravitation is much more than mere (apparent) attraction between matter-bodies. (Apparent) attraction between matter-bodies is relatively a minor aspect of gravitation. 'Hypothesis on MATTER' provides definite cause and specific mechanism for gravitational actions. Gravitation is the originator of all other actions in nature. All natural forces are derived from it. Various natural forces are nothing but different manifestations of only one type of '*force*' in nature. Gravitational actions create and sustain three-dimensional matter particles and macro bodies, constituted by them. Gravitation develops and sustains the universe, as we observe it.

While accepting existing theories, even though we knew them to be perfectly irrational, we had no hesitation to accept many assumptions like; 'actions at a distance', 'instantaneous action of gravitation, 'mass-less particles like gravitons', etc. to understand the phenomenon of gravitation. We bear with far too many assumptions, on which contemporary theories are based. Yet they do not give us a comprehensive theory of gravitation, encompassing all actions under all conditions. Hence, I request the reader to bear with the assumption of only one type of matter particles for the development of a comprehensive theory that can logically explain all phenomena related to matter, under all conditions, by the same basic laws. There is no singularity in nature, when physical laws are invalid. Applications of physical laws are identical at all places and under all conditions.

Since the book 'Hypothesis on MATTER' deals with all phenomena related to matter, it is brief in details and very elaborate in contents for those who are interested only in certain phenomenon. This book 'Gravitation' is compiled from explanations given in the book, 'Hypothesis on MATTER'. Only those details, directly pertaining to gravitation are included here. Other conclusions, reflected in this book, are from the parent book. Reader may kindly refer to the same for details of mechanisms of various phenomena, omitted in this book.

All criticisms, suggestions for improvement and enquiries are most welcome. Kindly address them to:

Nainan K. Varghese, Thiruvinal House, Adoor.P.O, Kerala State, 691523, India.

or mailto:matterdoc@gmail.com

Copies of the book "Hypothesis on MATTER" (second edition) are available at; http://www.booksurge.com and at many other on-line book stores.

I would like to express my heartfelt gratitude to Josey John, M.Tech. and Anita Varghese, without whose help and encouragement, this project could not reach this stage.

* ** *** ** *

GLOSSARY

2D energy field: Two-dimensional latticework structure, made up of quanta of matter, in a spatial plane.

Biton: A primary matter particle constituted by two photons.

Cooling: A process of raising the matter and energy content levels of a macro body.

Deuteron: A fundamental particle, constituted by many tetrons in the form of two spherical shells about one common positron.

Distortion: Geometrical deformation of 2D energy field-latticework squares.

Distortion field: A three dimensional region in 2D energy fields, where distortions are created by unstable photons.

Disturbance: A matter-body, formed by random collection of free quanta of matter.

Electric field: An angular distortion field.

Electric force: Reactive effort in 2D energy fields, produced by interaction between electric fields.

Electron: A hexton that has prominent south magnetic polarity and repulsive nuclear field in addition to an electric field.

Electromagnetic wave: Transmission of cyclically varying (in magnitude and direction) 2D energy field-distortions in space.

Energy: Stress, caused by distortions in 2D energy fields.

Field force: Effort exerted by distorted 2D energy fields.

Force: A functional entity that shows the rate of work, invested about a matter-body.

Free body: A matter-body that is under no other external influences, other than those considered (apparent gravitational attraction towards similar bodies in space).

Free space: A region in 2D energy fields, which has no distortions or disturbances, other than those considered.

Frequency: Rate of spin motion of a photon. Rate of cyclic variations.

Gravitational field: A distortion field created by a discontinuity in 2D energy fields, due to existence of a matter-body.

Halo: Outer fringes of a spinning galaxy, occupied entirely by free bitons.

Heat: A functional term to indicate the state of matter content level of a body.

Heating: A process of lowering the matter and energy content levels of a macro body.

Hexton: A fundamental particle, constituted by three bitons.

Inertia: Property of 2D energy fields that delays the completion of actions.

Inertial action: An action that invokes the property of inertia.

Inertial force: Reactive effort in 2D energy fields, produced by transfer of distortions from one matter field to another.

Inertial pocket: A three-dimensional region in 2D energy fields about a stable photon, where distortions required for the integrity and stability of the photon exist.

Latticework: Web-like structure formed by end-to-end linking by quanta of matter.

Latticework square: A rectangular section of latticework formed by four quanta of matter, as its sides.

Magnetic field: A linear distortion field.

Magnetic force: Reactive effort in the 2D energy fields, produced by interaction between magnetic fields.

Matter: A real substance that has positive existence in space.

Matter content: Quantity of 3D matter in a matter-body.

Matter content level: Average quantity of 3D matter in constituent photons of a macro body.

Matter density: Relation between total matter content and the volume of a matter-body.

Matter field: A three dimensional region in 2D energy fields about a 3D matter-body, where distortions required to maintain its integrity and state of existence is present. All additional distortions, required to sustain a body in its constant state of motion is also stored in the matter field.

Nuclear field: A radial distortion field.

Nuclear force: Reactive effort in 2D energy fields, produced by interaction between nuclear fields.

Neutron: A fundamental particle, made up of many tetrons in the form of spherical shell.

Photon: Corpuscle of radiation of matter (e.g, light). It is a compressed body of quanta of matter, sustained by gravitational pressure applied on it by the 2D energy fields.

Positron: A hexton that has prominent north magnetic polarity and attractive nuclear field in addition to an electric field.

Proton: A fundamental particle, made up of one positron and many tetrons in the form of a spherical shell.

Primary electric field: Angular distortion field about a biton.

Quantum of matter: Smallest independent matter particle (postulated in this concept).

Radiation: Transmission of 3D matter in the form of photons through the space.

Reaction: Effort produced by distorted 2D energy field, during its unstable state and stabilization.

Room temperature: Matter content level of matter-bodies, corresponding to the state of 2D energy fields in the surroundings of a large macro body.

GLOSSARY

Secondary electric field: Angular distortion field about an atom due the spin motion of atom's nucleus.

Space: A functional entity, presupposed by rational beings to represent the region of existence for matter bodies.

Temperature: Matter content level of a macro body, measured in terms of physical changes in another (reference) macro body.

Tetron: A primary matter particle constituted by combination of two bitons.

Time: A functional entity that shows the relation between displacement and rate of displacement.

Unstable photon: A photon, unstable by its inertial pocket, while remaining stable by its matter content.

Work: Magnitude of distortions in the 2D energy fields about a matter-body (in its matter field), required to sustain the matter body's integrity and state of constant motion.

Zilch force distance: Distance at which interaction between angular distortion fields do not produce inertial action.

* ** *** ** *

TABLE OF CONTENTS

FOREWORD
GLOSSARY

Chapter ONE
INTRODUCTION

1.1. Perspective of Gravitation: 15
1.2. Hypothesis on MATTER: 19

Chapter TWO
2D ENERGY FIELDS

2.1. Space: 23
    2.1.1. Aether drag: 27
    2.1.2. Perspective of Universal medium: 28
2.2. Quantum of Matter: 30
    2.2.1. Properties of quanta of matter: 33
    2.2.2. Formation of junction points: 35
2.3. 2D energy field: 40
    2.3.1. Reaction: 47
    2.3.2. Work and force: 47
    2.3.3. Self-sustenance of 2D energy field: 50
    2.3.4. Field force: 54
    2.3.5. Stabilization of 2D energy field: 55
    2.3.6. Equilibrium of 2D energy field: 57
    2.3.7. Properties of 2D energy fields: 62
2.4. Production of a Disturbance: 65
    2.4.1. Development of a disturbance: 68
    2.4.2. Magnitude of a disturbance: 69

Chapter Three
GRAVITATION IN 2D SPACE

3.1. Gravitation: 71
    3.1.1. Range of gravitation: 72
    3.1.2. Nature of gravitation: 73
    3.1.3. Strength of gravitation: 75

3.1.4. Gravitation on a point-disturbance: 76
3.2. Gravitation on a 2D disturbance: 78
    3.2.1. Shaping up a disturbance: 79
    3.2.2. Application of gravitation: 81
3.3. Action of gravitation: 83
    3.3.1. Motion of a particle by gravitation: 88
    3.3.2. Pressure energy of a disturbance: 90
    3.3.3. Gravitation on a straight perimeter: 91
    3.3.4. Gravitation on a curved perimeter: 95
3.4. Apparent gravitational attraction: 98
    3.4.1. Magnitude of apparent gravitational attraction: 101
    3.4.2. Effect of angle subtended: 105
    3.4.3. Many bodies in the same plane: 109

Chapter FOUR
CREATION OF 3D MATTER
4.1. Size reduction of a disturbance: 113
    4.1.1. Contraction of a small disturbance: 114
    4.1.2. Contraction of larger disturbance: 117
    4.1.3. Internal pressure of a disturbance: 118
    4.1.4. Very large disturbance: 120
    4.1.5. Disturbance of optimum size: 121
4.2. Creation of three-dimensional matter: 122
    4.2.1. Creation of higher-dimensional matter: 125
    4.2.2. Critical radial size of a 3D disturbance: 126
    4.2.3. Shaping up of a 3D disturbance: 127
4.3. Development of 3D matter particle: 128
    4.3.1. Ejection force: 129
    4.3.2. Spinning force: 131
    4.3.3. Ejection and spin of a disturbance: 132
    4.3.4. Centrifugal force in a disturbance: 134
4.4. Photon: 135
    4.4.1. Shape of a photon: 137
    4.4.2. Concepts of a photon: 139
4.5. Distortion fields: 142
    4.5.1. Linear distortion field: 145
    4.5.2. Angular distortion field: 146
    4.5.3. Radial distortion field: 147

4.6. Macro Bodies: 147
    4.6.1. Matter field: 150
    4.6.2. Motion of macro bodies: 152
    4.6.3. Inertia of rotary motion: 159
    4.6.4. Motion due to torque: 161
        Action of a rotating force: 163
        Action of linear force on rotating body: 166

Chapter FIVE
GRAVITATION IN 3D SPACE

5.1. Gravitation in 3D space: 170
    5.1.1. Apparent gravitational attraction in 3D space: 172
    5.1.2. Push gravitation: 173
    5.1.3. Gravitational attraction between photons: 175
        Gravitational attraction between coplanar photons: 179
    5.1.4. Macro bodies' apparent gravitational attraction: 185
    5.1.5. Magnitude of apparent gravitational attraction: 189
5.2. 3D system's gravitational constant in 2D system: 191
    5.2.1. Practical gravitational constant: 198
5.3. Action at a distance: 201
5.4. Screening the gravitation: 202
    5.4.1. Levitation: 204

Chapter SIX
MAGNITUDE OF ATTRACTION

6.1. Centre of gravity: 207
    6.1.1. Inverse square law: 211
    6.1.2. Breakdown of inverse square law: 213
6.2. Anomalies in gravitational attraction: 217
6.3. Mass and weight: 220
    6.3.1. Mass of a body: 220
    6.3.2. Weight of a body: 225
    6.3.3. Matter contents of fundamental particles: 228

Chapter SEVEN
GENERAL

7.1. Aether drag: 231
7.2. Heat and gravitation: 232

TABLE OF CONTENTS

- 7.2.1. Heating: 232
- 7.2.2. Acceleration due to external force: 234
- 7.2.3. Acceleration due to gravity: 234
- 7.3. Central force: 237
  - 7.3.1. Magnitude of radial velocity: 248
- 7.4. Planetary orbits: 251
  - 7.4.1. Linear motion of a rotating body: 251
  - 7.4.2. Orbital motion: 254
    - Inner solar System: 262
  - 7.4.3. Circular orbit: 265
  - 7.4.4. Elliptical Orbit: 269
    - Limits of angular speed during entry: 275
    - Orbits about a moving central body: 277
  - 7.4.5. Anomalies in planetary orbits: 281
    - Apparent loss of orbital motion of a planet: 281
    - Precession due to eccentricity: 282
    - Precession due to central body's path: 283
    - Perturbations caused by collisions: 285
  - 7.4.6. Electronic orbits: 285
- 7.5. Planetary spin: 286
  - 7.5.1. Spin due to central force: 288
  - 7.5.2. Unequal spin motion of a planetary body: 296
  - 7.5.3. Apparent spin motion: 297
  - 7.5.4. Anomalies: 300
  - 7.5.5. Variations in a terrestrial day: 301
- 7.6. Tides: 302
  - 7.6.1. Terrestrial tides: 316
  - 7.6.2. Direction of tides: 319
    - Magnitude of angular shift: 320
    - Direction of angular shift: 323
    - Direction of Solar tides: 328
    - Direction of lunar tides: 331
    - Effect of orbital motion on direction of tides: 332
- 7.7. Galactic repulsion: 334
- 7.8. Gravitational collapse of large bodies 341
- **Index** 347

* ** *** ** *

Chapter ONE
INTRODUCTION

1.1. Perspective of Gravitation:

Contemporary theories on gravitation, deal only with its dynamic actions on 3D matter-bodies. Dynamic actions due to gravitation are continuation of its actions on static matter-bodies. Static part of gravitational action is usually ignored in most theories on gravitation. Hence, gravitation is generally understood as a force (trying to or) moving a three-dimensional macro matter-body towards another. There are no accepted theories on its basic nature or mechanism of its actions in general (or on static bodies). I hope the concept, put forward the book 'Hypothesis on MATTER' can fill this vacuum and provide logical description to the nature and mechanism of gravitation in its static and dynamical states. According to this concept, gravitational effort or manifestation of its (force) actions is enormously stronger, when compared to other manifestations of natural efforts or actions (forces). All natural actions have their origin in gravitational actions. (Apparent) attraction due to gravitation between three-dimensional matter-bodies (the only effect of gravitation, known to present-day science) is relatively a minor by-product of gravitation in nature.

Currently, natural forces are assumed to be of distinctly different types; gravitational, electromagnetic and (strong & weak) nuclear forces. In mechanics, gravitation is the assumed universal (apparent) attraction between matter-bodies. It is assumed a separate kind from other types of natural forces. Although the gravitational actions do much more, apparent attraction between matter-bodies is the only action currently attributed to gravitation. No cause or mechanism of action are postulated.

Primarily, real gravitational actions in nature; create, sustain and destroy basic 3D matter particles and superior matter-bodies in space, under appropriate conditions. Apparent attraction between matter-bodies is in fact a by-product of separate gravitational actions on different matter-bodies. Appearance of attraction between matter-bodies is the result of separate and simultaneous gravitational actions on the matter-bodies. Apparent gravitational attraction between matter-bodies is a very minute fraction of gravitational actions (forces) on these bodies. It is because of this, that the gravitational force is incorrectly assumed the weakest of all natural forces. Contrary to truth, gravitational force is assumed to play no role in determining the internal properties of everyday matter-bodies. However, due to its long reach and universality, gravitational actions are assumed to shape the structure and evolution of stars, galaxies and the entire universe. Laws of gravitation determine the trajectories of bodies in the solar and galactic systems.

Works of Isaac Newton and Albert Einstein dominate the development of current gravitational theories. Building on the works of many great men leading up to Galileo (Galilei) and Johannes Kepler, Isaac Newton developed the first quantitative theory of gravitation. Newton's classical theory of gravitational force held absolute sway until Einstein's work in the early 20th century. Even today, Newton's theory is of sufficient accuracy for all but the most precise applications. Newtonian theory of gravitation is based on an assumed force, acting between all pairs of matter-bodies through empty space, i.e., an action at a distance. When a matter-body (mass) moves or changes its parameters, this assumed force, acting on other matter-bodies has been considered to adjust instantaneously to the new location of the displaced body and change its magnitude corresponding to the present parameters of matter-bodies. Without giving any reasons or mechanism of action, Newton held that every particle of matter in the universe attracts every other particle with a force that is proportional to the product of their masses (representing their matter contents) and inversely proportional to the square of distance between their centers (of gravity). This assumption was the result of observation of nature. Newton (or any other scientist) provided no reason or logical explanation for such (assumed) attraction between matter-bodies across empty space. This assumption formed the basis of all further developments of theories on gravitation.

Another assumption, prevailing at that time (and even today) is that planets orbit around their central bodies. This assumption played a crucial role in establishing the current theories on gravitation. It is from these closed geometrical figures of planetary orbits around the central body that proofs of contemporary gravitational laws were derived. Since no free body (a free body is under no other influence other than the effort considered) can orbit in closed geometrical path around another moving body, above-given assumption is baseless. Newton was able to show Kepler's three empirical laws of planetary motion mathematically, from his own three general laws of motion and the above mentioned law of gravitation. Power of these laws to explain and predict various phenomena were confirmed later. This made Newton's laws of gravitation infallible truths of science, all the while forgetting that the mathematical treatments used for the purpose are the (observed paths) apparent planetary orbital motions around a central body and not the true orbital paths of the planets about their central body.

We must consider that Kepler's laws of planetary motions were formulated at a time, when the phenomenon of gravitation or central force between earth and sun were unknown. What Kepler has done is to formulate laws to suite the observed locations of planets about the sun, which was considered to be static in space. No interactions or forces between central body and the planets were even considered as a cause of these relative motions of planets. Moreover, orbital path of moon was discarded. Probably, due to the realization that the moon could not execute an elliptical orbit around the moving earth. His laws are applicable only to the observed orbits of planets around a (static) sun. Despite these facts, Kepler's laws on planetary motion are routinely used in conjunction with many types of multi-body problems. Although mathematical treatments of apparent actions may produce results, which suit apparent phenomena, they cannot describe real facts.

Another but similar assumption is about the cause and nature of terrestrial tides. In mechanics, two linear forces acting on a body produce a single resultant action on the body. Yet two separate central forces on earth cause two separate and simultaneous sets of tides on earth. Smaller central force due to the moon causes larger set of tides, while greater central force towards the sun causes smaller set of tides. None of the exotic theories, formulated to justify these anomalies is logical. Centre of gravity of a macro body is an important assumption in Newtonian gravitational theory. Each macro body has a single centre of gravity. If more than one center of gravity is assigned to a single macro body, it is corrupting the phenomenon of centre of gravity. It is not correct to assume that an external macro body can gravitationally interact with different parts of a macro body, differently, as is considered in explanations on terrestrial tides.

Though, theoretically, vector equations in Newton's theory of gravitation could be used for finding the resulting gravitational field, when used for irregular non-spherical macro bodies, these equations are inefficient. In order to overcome this difficulty, potential theory was developed by scientists to deal with mysterious gravitational fields, mathematically. Potential theory led to the formulation of gravitational acceleration at any point in space called the gravitational potential. These are very useful in mathematical operations. However, at present, no one has clear understanding of what is a gravitational field.

Newton's conception and quantification of gravitation held absolute until the notion of instantaneous action at a distance (and through empty space), which his concept entailed, was recognized generally as unintelligible and illogical. In his general theory of relativity, Albert Einstein developed a wholly new concept of gravitation. Einstein proposed the idea of four-dimensional space-time continuum, which is curved by the presence of matter. This concept produces a universe in which bodies travel in geodesics (shortest paths). Geodesics replace the curved planetary orbits interpreted by Newton, as the result of some attractive force. Even in this case the planetary orbital paths are considered as closed geometrical figures around central bodies. Structure of space or the ability of a form-less space to bend or the mechanism of its wrap are not explained. Instead, non-physical space is assumed to possess physical characteristics to suit the theory.

Both, the above theories, rely heavily on the elliptical nature of planetary orbits around central bodies for their validity. Elliptical or circular natures of planetary orbital path are relative or apparent paths, as observed about the planets and their central bodies by observers on them. Kepler's laws of planetary motion are with respect to imaginary orbital paths of planets around a central body. Theories, based on imaginary concepts cannot be true. These theories also do not provide any explanations to the instantaneous nature of gravitational action or logically explain the mechanism of actions at a distance.

Special relativity theory considers that no physical signal travels faster than the speed of light and that all signals travel at this speed through empty space. No reason as to why or how they do so is given. Special relativity theory, with the field theories of electrical and magnetic phenomena, has met with much empirical success that most modern gravitational theories are constructed as field theories, consistent with the principles of special relativity. In a field theory, it is assumed that mere presence of a macro body produces a gravitational field, which permeates all surrounding space and become weaker in strength as distance from the body increases. A second macro body within this gravitational field is then acted upon by its gravitational field and experiences a force. This Newtonian force is then viewed as the response by the first macro body to the gravitational field produced by the second macro body. However, there are no explanations on the structure, constituents or the mechanism of development of these 'gravitational' fields. Often, the space itself is assumed to replace these fields.

Predictions of several phenomena by the relativistic view of gravitation violate Newtonian theory of gravitation and to the limits of observational accuracy; they have been confirmed by various experiments. However, prediction of relativistic (gravitation) theory about the existence of gravity waves, propagated by objects (matter-bodies), moving in a gravitational field or any implied existence of gravitational particles are not confirmed or detected yet. Moreover, the relativity theory has no logical explanation to the instantaneous gravitational actions or to the 'actions at a distance', without numerous assumptions.

Contemporary gravitational theories do not acknowledge gravitational actions on static bodies. Force, which comes into existence only during motion, is its sole expression.

1.2. Hypothesis on MATTER:

All diverse substances, including fundamental particles, in the universe are essentially made from the same basic material and in a similar manner. They should all be (basically) identical and obey the same basic laws under all conditions throughout universe. An alternative concept, that follows these notions are explained in the book 'Hypothesis on MATTER'. It primarily gives conceptual explanations of physical phenomena related to matter, with scanty mathematical treatments. All explanations in the book are with respect to a steady (absolute) external reference frame. 'Cause and effect relation' is strictly maintained in all explanations. By postulating an ideal 'basic matter particle' (the quantum of matter), it develops to explain all physical phenomena related to matter at all levels of complexity. Characteristic properties of postulated 'basic matter particle – quantum of matter' are so chosen that they are able to account for all physical properties of three-dimensional matter-bodies, space and all their apparent interactions in nature.

Save for the 'quanta of matter', no other bodies (virtual or otherwise) or characteristic properties are assumed in this concept. Quanta of matter are postulated with definite contents, structure and properties. They constitute all other superior matter-bodies. We come across three-dimensional matter-bodies in nature. 3D matter-bodies are developed through various steps of conversions of quanta of matter (without any changes to their fundamental nature) into more and more complicated and self-sustaining objects. These include different types of fundamental particles and larger macro bodies, found in nature. These self-sustaining objects (despite the fact that they are made from identical basic matter particles and in a similar manner), depending on their relative structure within the macro bodies, exhibit diverse properties and apparently interact with each other in different ways.

This concept has neither 'actions at a distance' nor the notion of 'pull' forces. All efforts of actions (forces) and their resultants are of 'push' nature. Forces, apparently, between matter-bodies are originated from and act through an all-encompassing medium. [However, in order to make the explanations clearer, unless specifically mentioned, present convention of forces being of 'pull' nature is also used in this book. Difference is that 'pull' forces are represented by arrows away from the point of application and the body is presumed to follow the point of action, in its motion. In case of 'push' forces, arrows representing the forces are directed towards the point of application and the body precedes the point of action]. As far as possible, conventional names of particles and effects are used (though some times they do not mean exactly the same) in this text. All such names are to be understood only in the context of this book. Figures, used in this text are not to scale. They are descriptive in nature and drawn in such a way as to illustrate the phenomenon presented. Term 'force' is often used in general sense to mean an effort that causes an action. Space, universal medium and 2D energy fields are synonymous.

Basic idea, presented in 'Hypothesis on MATTER', is nothing new. When 'Christiaan Huyghens' proposed the wave theory of light, it was assumed that light is propagated in the form of waves through an all-pervading hypothetical medium called 'aether' or 'ether'. As the theory advanced, other forms of radiations were also included with light, in the common heading of electromagnetic waves. Still later, as the existence of the aether could not be proved by experiments, notion of a propagating medium was lost. Nevertheless, the idea of radiation being in the form of waves is still maintained. At present, it is assumed that these waves of radiations are propagated through empty space at definite speed and no material-continuity of the space is needed for the transmission of the electromagnetic waves. This is clearly not logical. In this concept, it is seen that an all-pervading medium does exist and (in order to distinguish it from many types of conventional aethers) it is called by the name '2D energy fields' instead of aether. Considering the '2D energy fields', in place of the hypothetical medium 'aether' – although the medium is only a functional entity but no more a hypothetical one – the all-encompassing medium is given specific characteristic properties, corresponding to its constituents. By having a medium of propagation for all wave-like radiations, such phenomena become more logical. Having an all-encompassing medium does away with 'actions at a distance'. All other ideas expressed in the concept are developed from this basic idea.

Having an all-encompassing medium simplifies logical explanations on every phenomenon, related to matter. It provides continuity from one three-dimensional matter particle to another. There is no mystery for an action on one matter-body to affect another. All three-dimensional matter particles are immersed in 2D energy fields. 2D energy fields act on each of the 3D matter particles separately. These actions, when viewed in combination, produce the effects of apparent interaction between three-dimensional matter particles/bodies. Although the 2D energy fields are infinitely vast, they remain under compressed state. They have an inherent tendency to grow into any gap in them and tend to maintain their continuity at all times. It is this tendency of the 2D energy fields, which creates the phenomenon of gravitation.

According to the concept in 'Hypothesis on MATTER', gravitation is the result of two-dimensional latticework formations by the quanta of matter. It is the natural extension of inherent properties of quanta of matter, postulated in this concept as the most fundamental matter particles. Apparent attraction due to gravitation between matter-bodies (the only property of gravitation presently known) is relatively a minor realization of phenomenon of gravitation. Apparent attraction due to gravitation helps to sustain the universe and all its constituents in their current stable state. However, the unrealized part of phenomenon of gravitation is the basis of universe's existence. It causes the creation and sustenance of 3D matter, which makes up the universe and its constituent matter-bodies. Gravitation also originates all other manifestations of natural forces.

* ** *** ** *

Chapter TWO
2D ENERGY FIELDS

The concept, put forward in the book 'Hypothesis on MATTER', is based on the assumption that infinite number of 'quanta of matter' fills the entire space. The concept concludes that the universe (on an average) is eternal in its present state. Quanta of matter are the only postulated particles in this concept and they form the foundation of this concept. Properties of a quantum of matter are so chosen that they are able to account for all observed physical properties of matter, without help from imaginary particles or assumed properties. Quanta of matter, by their inherent properties, form latticework structures in each of the planes in space. These 2D latticework structures in all possible planes in space, together, forms the universal medium called '2D energy fields'.

The term 'force' is generally used to indicate an effort that causes an action. For specific actions, it indicates the rate of work-done, with respect to acceleration of a body. Free state of an object (a free body) denotes that the object is under no other external influence, other than the one that is considered. 1D, 2D and 3D indicate spatial one-dimensional, two-dimensional and three-dimensional status of real entities, respectively. Free space indicates a region in space; where there are no 3D objects or structural deformations present in the 2D energy fields, other than the objects considered and deformations associated with them. All motions/ displacements of matter-bodies are with respect to an absolute reference, which is defined later in this text.

2.1. Space:

Space is treated differently in different aspects. In physics; the space is understood as the boundless three-dimensional extent of universe, where all material objects (including the organisms and rational beings like ourselves) exist and in which objects and events occur. All material objects in the universe have their relative position and motion in the space. Space, itself, has no material existence. It is a functional entity that serves the purpose of locating various material bodies in it. Rational beings relate them with each other in space. The extent, outside material bodies, becomes the space.

Perception is a process by which living organisms become aware of relative positions of objects around them (and of their own bodies). For perception, living organisms use data received by their senses to conjure their own version of surroundings. This helps their orientation and activities with respect to their surroundings. It aids individuals to understand their location in relation to any other objects, with respect to depth, distance, etc. These are important for accounting their various motions. In order to be perceived, an object has to real, i.e. it should have positive (real) existence in space. With respect to three-dimensional rational beings, only 3D matter-bodies can have positive existence. Since the space has no material existence, it is a functional entity that is visualized by the rational beings for a purpose, assigned to it by them. Space has no real form or structure. A body that has no form or structure cannot deform or distort. Curvatures, expansion or contraction, etc. of (structure-less) space, used in some physical theories, are pure imagination which may aid mathematical exercises to prove illogical and mysterious physical laws.

All spatial concepts are related to contact-experiences of solid bodies (which may be felt by rational living beings). These include all those bodies, from which sensory perception is possible, under appropriate conditions. This has made it necessary to envisage an entity, independent of the (solid) bodies and yet embodying their locations. This functional entity outside material bodies, yet enclosing them came to be understood as space. Sole function of space is to provide material bodies with places for their existence. When a rational mind envisages a real object, it logically pre-supposes a place for its existence. This is understood not by sensing such a place but by the necessity of a place for the existence for any real body. This does not happen in case of functional entities, like: emotions. In this sense, the space appears to have a physical reality, which solely depends on the existence of (real) material objects in it. As a result, the notion of space is somewhat incoherent, because it professes to be a container that is logically prior to its contents. Space turns out, in practice, to be merely an indefinitely extensible collection of its contents – the 3D matter-bodies. Everything that occupies space falls within this wider spatial context. Space denotes a property by virtue of which different bodies occupy different positions in the universe. The possibility of arranging an unlimited number of matter-bodies next to one another denotes that the space is infinite in its extent.

There is no logical argument for theories based on concepts of imaginary entities like space. That is why, from early time, it was believed that an entity, named aether, filled the entire space. In these theories, the aether replaced the space by filling it entirely. Therefore, all properties assigned to space could be the properties of aether. Aether had an ambiguous form but it was regarded as a real entity. Since the aether was real, it could deform, move or otherwise interact with other material objects. Unfortunately, no one could describe a satisfactory structure or properties for the aether. Many types of aether were proposed, as required for various theories. Aether was assumed to be weightless, transparent, frictionless, undetectable chemically or physically, and literally permeating all matter and space. Aether-theories met with increasing difficulties as the nature of light and the structure of matter became better defined, even if it was on imaginary basis. Since there is no accepted definition of aether, scientists concentrated their effort to find an effect, the aether may make on other 3D macro bodies. For this, they assumed, when a large macro body moves through aether, the macro body should essentially experience a drag due to the friction between the two. Aether-theories were seriously weakened (1881) by 'Michelson-Morley experiment', which was designed specifically to detect resistance to the motion of Earth through the ether. Experiments showed that there was no such tangible effect. Finally, when aether's existence could not be proved experimentally, by experiments based on illogical theories, majority of scientists abandoned the concept of aether. They have turned to more mysterious concepts of space.

Everyday experience of natural phenomena shows mechanical things are moved by contact between force-applying body and force-receiving body. Thus, we came to conclude that for any action to take place between two real bodies there must be a contact between them. Nature of this contact is expressed as action of force between the bodies. Any 'cause and effect' without a discernable contact between participating bodies (or 'action at a distance') contradicts common sense and has been an unacceptable notion since earliest of time. Whenever the nature of transmission of certain actions and effects over a distance was not understood, even today, as a conceptual solution of transmitting medium, aether (in the form of various types of fields) is resorted to. However, any description on how the aether, in the disguise of different fields, functions remains vague. But its existence in the form of various fields was required by common sense and thus not questioned. Aether (expressed as various types of fields) was discovered during the heyday of aether theories, according to which all space is permeated by a medium capable of transmitting forces between 3D matter particles. The electric and magnetic fields were interpreted as descriptions of different states of strain of the aether, so that the location of stored energy in space was like as it would be in a compressed spring. With the abandonment of the aether-theories, following the rise of relativity theory, this imaginary model ceased to have validity. However, the original aether is preserved by us in the form of various fields in our theories. This is because, an all-encompassing universal medium is essential to destroy the myth of 'action at a distance through empty space', which is the worst illogical assumption of modern science. There are many forms of fields, used in various theories, each one proposing different types of fields with vague properties of aether.

Space is also viewed only as a conception. Since the space provides an extent for real or 3D material bodies to exist, the concept of '3D material object' is necessary to define space. The concept of '3D material object' is linked to our sense-experiences, which continue through certain time. Existence of real objects is thus of a conceptual nature, linked to our sense-experiences. Existence or reality of material bodies are defined simply as concepts of our mind, which depends wholly on their being connected with our sense-experiences. Argument, supporting these types of theories, is that a rational being's thoughts and concepts are created by experiences of his senses with 3D material objects. These experiences are meaningful only with reference to his senses.

His thoughts are products of his mind's activity. As long as the mind can act in certain way, existence or reality of objects is immaterial to understand his surroundings. Therefore, no wise logical consequences of the sense-experiences are required to understand the universe or actions in it. Although this argument overlooks that the presence of real objects is necessary to produce sense-experiences, without which mind's activity cannot take place, it is very useful to produce exotic and mysterious physical theories. In these theories, the space is often linked with another functional entity, 'time', to form yet another functional entity called the 'space-time continuum'.

2.1.1. Aether drag:

It was the failure to notice the assumed aether drag on earth's motion through the space that ended the progress in the search for an all-encompassing universal medium. This was unnecessary, because the assumption of 'aether drag' itself is unwarranted. In this concept, explained in 'Hypothesis on MATTER', it is shown that every basic 3D matter particle is moved by a universal medium at the highest possible speed. Basic 3D matter particles (constituting primary particles, fundamental particles and superior matter-bodies) have their paths curved to limit the sizes of their circular paths to within the primary particles. A macro body consists of millions of basic 3D matter particles, moving in circular paths, within its body. Any motion of the macro body is achieved by simple displacements/deflections of basic 3D matter particles' circular paths in space. It is the universal medium that is affecting such motion. Matter has no ability to move on its own. Since the universal medium is the one, which is displacing macro body, there will be no relative motion or friction between them. Action is limited to the universal medium, present within and in the immediate neighbourhood of the macro body.

Motion of a matter-body, through the universal medium, is like the motion of a floating body in a narrow ocean current. Ocean current carries the floating body along with it and there is no relative motion between the floating body and the surrounding water. However, this floating body has a clear relative motion with respect to the vast ocean. Similarly it is the moving distortions in the universal medium, which moves a matter-body. This part of universal medium, which is engaged in moving a matter-body, is a local region of universal medium in and about the matter-body. Distortions in the universal medium carry the matter-body along with them and there is no relative motion or friction between the distortions and the matter-body. However, with respect to the vast universal medium outside the immediate surrounding of the matter-body, the matter-body has a relative displacement in space.

Basic 3D matter particles, during their motion through the universal medium, experience resistance from the universal medium. Basic 3D matter particle's ejection effort (moving force) is also caused by the universal medium. Speed of the basic 3D matter particle is determined by the resultant of these efforts. And this speed happens to be the highest possible speed of any kind of transmission through the universal medium that can be sustained without breakdown of universal medium. Since the resistance from universal medium is already accounted for, in the motion of the basic 3D matter particles, such resistance will not be carried further into the motion of 3D macro bodies. Therefore, 3D macro bodies will not experience any drag to their motion through a universal medium (space).

2.1.2. Perspective of Universal medium:

Rational thoughts suggest the existence of an all-encompassing medium, which fills the entire space, including inter-particle spaces within macro bodies. Media, suggested in various aether-theories, assumed that the medium, which fills the entire space, is of a different substance from matter and all matter-bodies are immersed in it. Actions were attributed to matter-bodies, whose actions would affect other matter-bodies through the medium. These assumptions required that the medium should have certain properties. These properties, when taken together, often contradicted themselves.

However, in many contemporary theories, we still use a vague form of aether – various types of fields. These fields have no particular constituents, structure or properties. Field, used in any particular theory, happens to behave as required by the theory.

Each type of field is associated with a particular phenomenon. Although they lack physical structure, they are capable to interfere or interact with each other to produce different types of physical actions. These fields are mainly used in analytical explanations to indicate the region of influence of a phenomenon, in space. Imaginary 'lines of forces', used in these fields, facilitate better mathematical understanding of a phenomenon. At times, even the formless entity of space is assigned with physical properties, so that it is capable of deformations.

We are 3D beings. All our actions and observations are limited to 3D matter-bodies and our sensory perceptions. A universal medium is neither a 3D matter-body nor it can cause sensory perceptions. Since we cannot perceive the universal medium, directly or through experiments, it is to be concluded that the universal medium exists outside the scope of 3D matter-bodies. Hence it is impossible for us to observe or act on the universal medium, directly. This does not preclude the existence of universal medium in other dimensional space systems (which are beyond the comprehension of 3D beings) or its actions on 3D matter-bodies. If all phenomena, related to matter can be logically explained by such a concept on an all-encompassing universal medium, it should be recognised as true.

All-encompassing medium, suggested in 'Hypothesis on MATTER', has a definite structure and properties. Although it is made up of real matter particles, it remains outside the scope of 3D matter-bodies. The medium fills the entire space outside 3D matter particles. It has the same matter density as the matter density of basic 3D matter particle, yet it remains outside our sensory perception and behaves like a perfect fluid to the relative motions of 3D macro matter-bodies. It causes no friction to moving matter (macro) bodies. It acts as a medium for all apparent interactions between 3D matter-bodies. Above all, it is this medium that creates basic 3D matter particles, out of disturbances in it, sustains them in stable states and in due course of time destroys them. The medium cannot interfere with any apparent actions of matter-bodies because this medium itself produces all such actions rather than the matter-bodies. Universal medium inherently seeks serenity in nature.

Chapter 2. 2D ENERGY FIELDS

2.2. Quantum of Matter:

'Quantum of matter' is postulated as the basis of this alternative concept. No other imaginary particles or assumed properties are envisaged in this concept. Observable universe owes its development and sustenance on the quanta of matter and their inherent properties, as specified in the postulation. Creation, development and apparent actions of 3D matter-bodies strictly follow 'cause and effect' relations. By following consistent reasoning, this concept is able to develop a theory that can logically explain all physical phenomena related to matter. Details on the properties and actions of quanta of matter are given in the parent book.

In brief: The postulated 'quantum of matter' is a real matter particle that has positive existence in space. Matter is its substance. A quantum of matter is a very small bit of matter. It has a definite structure, properties and ability to act. It has its existence in all spatial dimensions, however small such measurements may be. Each quantum of matter is an independent matter-body and it keeps its individuality under all conditions. However close they come to each other, matter content in one of them cannot amalgamate into the matter content of another. Quantum of matter or its matter content cannot be divided, destroyed or re-created. A quantum of matter, in its natural free state, is postulated as a single-dimensional object. It has only the length as its tangible measurement. Measurements in other two spatial dimensions are too small to be tangible. The quanta of matter, even in their 1D status, are real matter-bodies. For 3D beings, all real entities need to be three-dimensional. We, being 3D beings, are unable to appreciate quanta of matter's real existence in our 3D spatial system. Because of this difficulty, we may consider quanta of matter in their 1D and 2D states as functional entities.

A quantum of matter is an independent (real) matter-body. It has a measure of its matter content. Since we have no reference, matter content of a matter-body is not directly tangible. Matter contents of superior bodies are usually measured by their equivalents of 'rest mass' or 'weight'. Magnitudes of matter content in different quanta of matter may vary from one quantum of matter to another. Original matter content of a quantum of matter can neither be reduced nor increased. Natural tendency of 'self-elongation' in the first spatial dimension reduces a free quantum of matter's existence in all higher spatial dimensional systems to be negligibly small. Matter content of a quantum of matter tends to maintain integrity and individuality of the quantum of matter under all conditions. Due to the adhesive property of its matter content, a free quantum of matter tends to grow in single spatial dimension, while reducing its measurements in all other spatial dimensions. Even though, the dimensions of a quantum of matter in the spatial dimensions, other than its single spatial dimension, are negligibly small, a quantum of matter has positive existence in all spatial dimensions. Thus, a free quantum of matter is a 1D matter-body with positive existence in all (three) spatial dimensions.

Matter content (within a quantum of matter) has a tendency to coagulate together. This tendency preserves quantum of matter's independence. Matter content, at nearest points (within a quantum of matter or between different quanta of matter in direct contact in the same spatial dimension), has adhesive property (a tendency to merge). Ability to maintain individuality of a quantum of matter by the coagulation of matter content may be called the property of 'self-adhesion'. Due to the property of self-adhesion, the matter content of a quantum of matter continuously tries to coagulate together. Self-adhesion provides the matter content of a quantum of matter with a property similar to 'attraction' within itself. Self-adhesive nature of a quantum of matter tends to gather its matter content towards a central point. Property of self-adhesion gives a quantum of matter its two fundamental characters of, 'self-constriction' and 'self-elongation'.

Property of self-constriction compels a quantum of matter to restrict or reduce its existence in any dimensional space system to minimum. Self-elongation compels a quantum of matter to grow (revert) into lower dimensional space system from its present spatial state until it reaches single-dimensional state by lengthening-out as far as possible. These properties squeeze the matter content of a quantum of matter to lengthen-out along its axis. This is an inherent property of quanta of matter. A quantum of matter has a natural tendency to grow in its own single-spatial dimension. Reducing the length of a free quantum of matter by external means, against self-elongation, compels the quantum of matter to grow into second spatial dimension. Similarly, reducing the area of a 2D quantum of matter (a quantum of matter that is fully converted into 2D status) in its spatial plane compels it to grow into the third spatial dimension.

If left free, in free space, a quantum of matter will grow in length, indefinitely. As and when external actions reduce the size of a quantum of matter in the present dimensional space system, its body expands into higher dimensional systems. Depending on the external pressure on it, a quantum of matter is able to exist in single-dimensional, two-dimensional or three-dimensional spatial states. On compressing a 1D quantum along its axis, against self-elongating action, its length reduces and its body grows, against the self-constricting action, into 2D space system. Length of a quantum of matter is reduced by its body growing in 'width' to accommodate its matter content. Thus, it develops into the second-dimensional space until it becomes a circular object. Application of external pressure (forces from all around) reduces the area of a two-dimensional quantum of matter by its body growing in thickness and thus developing into the third-dimensional space.

A quantum of matter, being free in the space, where it can lengthen-out up to its limit is only a hypothetical consideration. As far as a single quantum of matter is concerned, there is no unlimited 'free' or 'empty' space. [Distance is a measure of separation between matter particles/bodies with respect to a reference. A single quantum in space has no other body of matter as a reference]. Since the entire space is filled with 2D energy fields, made up of quanta of matter, there can be very little free space available to any free quantum of matter.

By postulating the quanta of matter of definite matter contents, matter content of a composite body (a union of more than one quantum of matter) is quantified in terms of number of quanta of matter. Only an addition or subtraction of integral numbers of quanta of matter can change the total matter content of a composite body. Due to possible variations in their matter contents, between quanta of matter, magnitude of total matter content of a composite body may not be strictly proportional to the number of quanta of matter, the body contains.

All properties of quanta of matter, given above, are part of the postulation of the quanta of matter. Fundamental character of self-adhesion (of a quantum of matter) produces these properties. Every inference in this concept is based only on these properties of the quanta of matter. Basic postulations of this concept also include the assumption that the entire space is filled with quanta of matter in definite (self-created) structural formations. [Space is a functional entity that indicates the place of existence of quanta of matter. Space is created by the tendency of rational beings to presuppose a 'place of existence', when thinking about a matter-body. Space has no other functions. Since the space has no physical form or structure, it cannot be curved or deformed].

2.2.1. Properties of quanta of matter:

'Quantum of matter', postulated in 'Hypothesis on MATTER' has a vague resemblance to the 'strings' in 'string theories'. However, it should be noted that the quanta of matter are distinctly separate from the 'strings'. Only resemblance is that both of them are postulated basic entities. Generally, a 'string' is assumed to be a vibrating particle of energy in space, which forms various matter particles in accordance with the frequency of its vibrations. Energy is not defined and what is vibrating or how does it vibrate are left to the imagination. Unlike a string, a quantum of matter is a matter particle with positive existence in space. A quantum of matter may be transformed to exist in any of the dimensional space systems by (or lack of) external effort. Other than its inherent property to elongate in its own spatial dimension and the tendency to form one-dimensional chain with similar bodies in contact, it is inert and incapable of any other motion or interaction on its own. A quantum of matter has definite properties, quite different from those of 'strings'.

Properties of the quanta of matter, postulated as the most basic entity to form the foundation of this concept may be tabulated as follows:

1. Quanta of matter are the indivisible basic matter particles. They can neither be created nor destroyed. They have perpetual existence.

2. A quantum of matter, in its 1D state, has a body and two ends. It keeps its single-body identity under all conditions. Matter content of a quantum coagulates to be single body. This provides a quantum with its self-adhesive nature.

3. When free, a quantum of matter exists in its single-dimensional space system. Its body extends in its own dimensional space so that body's measurements in all other dimensional systems are negligible (reduced to zero). Self-constriction is an inherent nature of quanta of matter.

4. External compression, acting from the ends of a 1D quantum of matter, may reduce its length and compel it to grow into the second-dimensional space system and external compression, acting all around the perimeter of a 2D quantum of matter, may reduce its area and compel it to grow into the third-dimensional space system and so on.

5. Two quanta of matter, in contact, attract each other and they tend to line up in the same one-dimensional space system so that they form a chain in a straight line. Development of a chain of infinite length (with excess number of quanta of matter in each chain) keeps the quanta of matter in a quanta-chain under compressive pressure from their ends.

6. There are infinite numbers of quanta of matter in nature. They arrange themselves to form latticework structures, called '2D energy fields', in all planes and in all directions to fill up the entire space. A single quantum of matter of matter has no stable, independent existence in 3D space system.

7. Two quanta of matter, in their 1D and 2D states and intersecting at a point, are able to co-exist at that point without hindrance to their independence.

8. Dimensions of a quantum of matter are so small that it exists and acts as a functional entity with respect to 'rational 3D beings'

9. Matter contents of all quanta of matter need not be equal. Quanta of matter, which have nearly equal matter contents, form the 2D energy fields and others create disturbances in the 2D energy fields.

Quantum of matter (with definite properties and plentiful in nature) is the only postulated particle in this concept. Since it is a postulated particle, no questions on its creation or reasons for its existence arise. At present, we use many postulated particles and numerous assumed properties to describe various phenomena in science. There are too many postulations to give a logical explanation on any phenomenon. In this concept, explanations on all phenomena are based only on this one type of postulated particles.

2.2.2. Formation of junction points:

For explanations in this paragraph, quanta of matter of equal matter contents are considered. Quanta of matter, when in contact with each other, have a tendency to form one-dimensional quanta-chain extending infinitely. Details on the mechanism of interaction between quanta of matter can be found in the parent book. Due to the inherent tendency of quanta of matter to grow in length, theoretically, one or few quanta of matter can form a chain of infinite length. However, in nature, there are numerous occasions for each quanta-chain to be broken in the presence of readily available quanta of matter to migrate into the gap created by the breakage. Hence, each quanta-chain has excess number of quanta of matter in it. Excess number of quanta of matter and their linear alignment in a quanta-chain, compel the constituent quanta of matter to press on each other. Quanta of matter in a quanta-chain remain under axial compression, corresponding to the present state of universe.

In figure 2A, 'AO' and 'OB' are two quanta of matter in a one-dimensional quanta-chain. Let there be another quantum of matter, 'CO_2,' in contact with quantum of matter AO in the chain. As shown in the figures, lengths of quanta of matter in the chain are reduced and their widths are developed by the compression from the chain. The free quantum of matter CO_2 has no breadth but has only length as its tangible measurement. Attraction between matter contents of these quanta of matter, in contact, moves quantum of matter CO_2 until its end 'O_2' join the common junction 'O'. When such a common junction 'O', as shown in figure 2B, is formed, quanta of matter AO, BO and CO_2 will continue to apparently attract each other so that they may align in one-dimensional space system. Magnitude of apparent attraction between two adjacent quanta of matter at a junction is inversely proportional to the divergence angle between their axes. Disregarding the turning movements of quanta of matter AO and BO (on each other), apparent attraction turning quantum of matter CO_2 clockwise due to interaction between quanta of matter AO and CO_2 is greater than the apparent attraction between quanta of matter CO_2 and BO trying to turn quantum of matter CO_2 anti-clockwise. As a result, quantum of matter CO_2 will turn clockwise and settle so that its body is at equal angular difference from both the quanta of matter AO and BO. In this position, moving efforts (or torques) applied on quantum of matter CO_2 by both quanta of matter AO and BO of the quanta-chain are equal and opposite in directions. Hence, the efforts neutralize each other. If quanta of matter AO and BO of the chain keep their mutual alignment in straight line, the quantum of matter CO will come to settle in the second spatial dimension, at right angle to the chain formed by the quanta of matter AO and BO, as shown by quantum of matter CO_2 in figure 2B.

While considering the turning movements on quanta of matter AO and BO, due to the presence of quantum of matter CO_2, apparent attractions on quanta of matter AO and BO by their outer

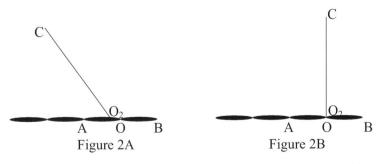

Figure 2A   Figure 2B

neighbors in the quanta-chain also will have to be taken into consideration. There will be small angular movements of quanta of matter AO and BO, balanced by the reactions from the chain. Quantum of matter CO_2 will remain at its full length until two other quanta of matter join at its ends in its own spatial dimension. When both ends of the quantum of matter CO_2 are linked to other quanta of matter in a quanta-chain in the same spatial dimension, apparent attraction between them will reduce the length of quantum of matter CO_2 also to the minimum limit and develop its negligible breadth.

Another quantum of matter (in the same plane), joining at the junction 'O', will line up itself, again at right angle to quanta of matter AO and BO, but in a straight line with quantum of matter CO_2. Such a junction of four quanta of matter makes the turning efforts on all participating quanta of matter about the junction 'O', in a neutral and stable condition. All four quanta of matter at the junction achieve stable states, when they are at right angle to each other, as shown in figure 2D. An angular diversion, of these quanta of matter, from their stable position is automatically corrected by the aligning stress between them. As long as unstable angular differences between the quanta of matter at a junction are maintained, these quanta of matter remain under strain to return to their stable position. This strain produces a continuous stress (effort) on the participating quanta of matter to return to their neutral and stable condition. In this case, their stable positions are mutually at 90° to each other.

Figure 2C shows a junction point, 'O', of four quanta of matter in latticework formation. In its original condition, quanta of matter A_1 (shown by grey dashed line), B, C and D are in their stable condition. We shall consider the turning efforts applied on quantum of matter A_1 by the quanta of matter B and C. Quantum of matter D, which is also a constituent of the junction, does not make direct contact with quantum of matter A_1. Therefore, there is no direct interaction between them. In their stable position, with quantum of matter OA_1 perpendicular to both quanta of matter BO and CO, interactions between quanta of matter A_1 & B and quanta of matter A_1 & C, acting on quantum of matter A_1 are of the magnitude 'F' each, shown by thick arrows in dotted lines. Since these efforts are of the same magnitude and opposite in directions, they neutralize each other. There is no resultant turning action on the quantum of matter A_1.

Let the quantum of matter A_1 be deflected by an angle F to position 'A'. Angular difference between quanta of matter A and C is an acute angle. Angular difference between quanta of matter A and C has reduced from 90° by an angle Φ. Interactive turning effort between quanta of matter A and C, applied on quantum of matter A increases correspondingly, to 'F_2', shown by thick black arrow. Angular difference between quanta of matter B and A has exceeded 90° by an angle Φ. Interactive turning effort between quanta of matter B and A applied on quantum of matter A decreases correspondingly, to become 'F_1', shown by thick black arrow.

Turning efforts 'F_2' and 'F_1' are in opposite directions. Resultant of these two efforts acts in the direction of greater effort 'F_2' and turn the quantum of matter 'A' back to its stable position at 'A_1'. Due to the deflection of quantum of matter 'A' from its stable position, similar realigning efforts will also develop between other quanta of matter at the junction. As a result, all quanta of matter at a junction have an inherent ability to align in the same plane with equal angular difference between adjacent quanta of matter.

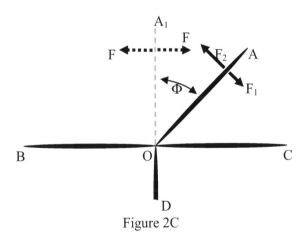

Figure 2C

Movements (displacements) of participating quanta of matter at a junction are 'work'. Stress at the junction, between the quanta of matter is always present, even in their stable state. It is their relative angular differences, which produces a resultant effort to move any or all quanta of matter at the junction. Rate of application of this effort is called 'force', when it is related to distance (acceleration) and it is called 'power', when it is related to time. Hence, we shall consider the stable condition as a reference state and consider only the resultant stress at the junctions for all our explanations. As long as the relative displacements (from their stable positions) between the participating quanta of matter at a junction are present, certain work is present in association with that junction of the quanta of matter. Work-done, at the junction, is maintained as long as quanta of matter at the junction remain displaced from their stable position. Displacement of quanta of matter or the work-done at a junction produces an associated stress between the participating quanta of matter. Stress, produced by the strain developed at the junction due to displacement of constituent quanta of matter is the 'energy' developed or stored at the junction. This stress is released and the energy stored is reduced to nil, when the participating quanta of matter return to their stable positions, with equal angular difference between adjacent quanta of matter. In this state, the work-done at the junction is undone.

Quanta of matter, meeting at a junction, being parts of dimensional quanta-chains of their own, are already (axially) compressed to their minimum length in the 1D space system. Minimum stable lengths of all quanta of matter in the chain-formation are equal and correspond to current state of universe. Total magnitude of compression in any part of a quanta-chain is proportional to the extent (length) of the chain. Compression in quanta-chains increases widths of constituent quanta of matter only by negligible values. Therefore, they remain within the definition of one-dimensional objects. They are at the brink of growing into the two-dimensional space system. Additional external compression on the quanta of matter, along their axes will make them to grow, for the duration of such compression, into the second-dimensional space system. Quanta of matter grow in width by reduction in their length. When the additional compressions from the ends are removed, the quanta of matter will return to their original state.

2.3. 2D energy field:

The second spatial dimension is common to all quanta of matter, meeting at a junction. A latticework of such arrangements of quanta of matter, as a whole, can operate only in its own two-dimensional space system. Deformation of the latticework (strain in the latticework) develops corresponding stress in the structure. This stress is understood as energy. As the deformation is removed the stress/energy disappears from the latticework structure. Since this latticework structure of quanta of matter stores and delivers the functional entity of energy, required for all inertial actions, it may be called a '2D energy field'. 2D energy fields are real entities made up of quanta of matter. Their deformed (unstable) states cause the development of functional entity of 'energy'. No quantum of matter can remain in the third dimensional space system (with respect to a 2D energy field in space) and link with the junctions of latticework of a 2D energy field. Each plane in space has an independent 2D energy field.

Any number of quanta of matter in the same plane may form a junction. Quanta of matter, forming a stable junction settle around the junction point, in the same plane, with equal angular difference between neighboring (adjacent) quanta of matter. However, junctions with four quanta of matter (neighboring quanta of matter perpendicular to each other) provide most stable configuration of a 2D energy field. For this, the quanta-chains in a plane settle perpendicular to each other and crossing at the junction points, to form a latticework. Each quantum of matter occupies one side of a square formed by the quanta of matter in the latticework. Only quanta of matter of (somewhat) equal matter contents make stable latticework by quanta-chains. Latticework structure, formed by the quanta of matter, is a 2D energy field. A 2D energy field extends infinitely in its plane, in all directions.

Let us consider a free quantum of matter, making contact with another quantum of matter, which is already a part of a junction or a quanta-chain. Although the quanta of matter at the junction are one-dimensional objects, they have certain negligible widths. On making body-contact, the free quantum of matter is influenced by other quanta of matter at the junction point, by the self-adhesive nature of their matter contents. Free-quantum of matter develops its negligible width in the same plane as that of the quanta of matter at the junction point in the chain. As a result, a latticework of quanta of matter, forming a 2D energy field, can exist only in one plane. It should also be noted that, in our three-dimensional system, a 2D object has no thickness. Any number of 2D objects stacked parallel and on top of the other, will not increase the thickness of the stack from being negligible or zero. This is true only as long as a plane is considered to have no thickness. However, in this concept, quanta of matter are real entities with positive existence in 3D space. Although negligible, with respect to three-dimensional space system, a plane (and the quantum of matter in its 1D and 2D states) has certain thickness.

Given enough quanta of matter and time, the quanta of matter will form 2D energy fields in every possible plane in space. Therefore, this concept assumes that the entire space is filled with infinite number of 2D energy fields. For this, all parallel planes also need to have their own 2D energy fields. Having 2D energy fields in all these parallel planes increases the thickness of space occupied by them. Consequently, it is logical to think that each 2D energy field and its plane have certain thickness. Similar logic is applied in case of the 2D energy fields in all other directions also.

Development of a 2D energy field is a natural process. 2D energy fields can be formed in all adjacent planes and all planes crossing each other at various angular directions. However, because of their independence and individuality, two 2D energy fields do not interact (directly) even if they are touching (or crossing) each other. Distance or separation between two adjacent and parallel 2D energy fields corresponds to the thickness of a 2D plane or of a 2D energy field in the 3D space system. Magnitude of this thickness has to be a positive value. A 2D plane with zero thickness or negative thickness is a non-entity. Independent 2D energy fields, extending to infinity, exist in all possible 2D planes of the 3D space. Therefore, the number of 2D energy fields in a given volume of 3D space cannot be ascertained with reference to the 3D space system.

A quantum of matter, being a single-dimensional body, it cannot bend into another spatial dimension. Although a 2D energy field is made up of rigid 1D quanta of matter, its latticework structure and flexible joints makes it distortable in its plane. Distortions of limited magnitude are tolerated within a 2D energy field's latticework structure. During distortions of a 2D energy field, two actions may take place: (1). Quanta of matter at a junction point are angularly deflected from their stable alignment with respect to each other and/or (2). Quanta of matter in the quanta-chains vary their length, depending on the variation of compressive pressure from their ends.

Angular displacements of quanta of matter at a junction point invoke reactive efforts on the quanta of matter to return to their stable positions. Similarly, a change in the length of a quantum of matter also invokes reactive effort in the latticework to restore its stable configuration. Any distortion in the 2D energy field is always opposed by a reaction. This reaction, from within itself, tends to restore the stability and serenity of a 2D energy field. Thus, it becomes an inherent property of the 2D energy field to strive towards its stable state. In its stable state, a 2D energy field is isotropic, homogeneous and serene. Each 2D energy field, considered as a whole, remains steady and perpetual in space. Small local distortions in it may develop or may be transferred within its plane. Hence, on a large scale, the 2D energy fields, being steady, can provide an 'absolute reference' in space.

Due to the latticework structure of a 2D energy field and its inherent property of stabilization, distortions in a 2D energy field cannot be preserved static in a locality. Any distortion is bound to spread-out in the latticework. If there is an external cause (in a sense that the cause may be anything other than the 2D energy field itself), the distortions tend to be transferred in the direction of creation of distortions by the cause. Sequential spread of distortion from one latticework square to the next, introduces a time delay in the development and transfer of the distortions from one place in the 2D energy field to another place in it. As soon as the cause is removed, latticework structure tends to regain its stability. However, the distortions, contained in the latticework, will continue to be transferred in its original direction, unless they are removed by an external agency, by introducing equal but opposite distortions in the latticework. Due to homogeny of a free 2D energy field, distortions in it are transferred at a constant speed through it. The property of time delay during the development and transfer of distortions and the constant speed of their transfer through the 2D energy field give rise to the property of 'inertia'. Inertia is a property of 2D energy fields.

A distorted region in a 2D energy field is a 'distortion field'. Depending on the type and direction of distortions, distortion fields can be classified into gravitational field, electric field, magnetic field, nuclear field or inertial field. This type of classification gives the present concepts of various 'fields' definite structure and real existence in space. Due to the latticework structure of the 2D energy field, distortions in it can exist only in a closed-loop arrangement. Every plane in space contains one 2D energy field each. 2D energy fields in different planes, passing through a point, co-exist. This is possible because of the ability of quanta of matter to cross each other at an angle without affecting each other. They can neither interact nor can the actions in one 2D energy field be transferred into another 2D energy field, directly.

Theoretically, displacements of quanta of matter (including the changes in their lengths) are tangible in 2D space system. They constitute 'work-done', within the distorted region of a 2D energy field. Stress, produced in the latticework structure of the 2D energy field, by the distortions in it, is the 'energy' associated with the work-done. Rate of distortions (work), being introduced into a region of a 2D energy field latticework (from another region of the same 2D energy field), is the 'force or power'. 3D matter particles are opaque to distortions in 2D energy fields. Transfer of distortions in the 2D energy fields affects any 3D matter particles present in the region by carrying them along with the distortions. Ultimately, transfer of latticework distortions from higher distortion-density region to lower distortion-density region produce displacements of matter-bodies (disturbances) in 2D energy fields. This is the action of an effort (generally called force).

Whichever is the manifestation of force (gravitational, electromagnetic, nuclear, inertial, etc.); they all act in similar manner and their mechanism of action is same. Thus, fundamentally, there is only one type of force in nature. Force is generally associated with motion of a 3D body and it simply means rate of work, irrespective of the nature of work or its source. We make distinction between various manifestations of force, by observing different phenomena in nature. It is futile to unify various man-made distinctions of forces, mathematically, without realizing the actual mechanism of action of force. Once the mechanism of action of a force is understood, these distinctions disappear and all types of forces and their actions become similar.

2D energy fields fill the entire space. Due to occupancy of a volumetric space by 2D energy fields in all planes passing through that space, the entire volume is filled by quanta of matter. Total matter content within this volume of space is comparable with that of a 3D matter particle occupying the same volume of space. Since the 2D energy fields cannot act among themselves, matter content enclosed within this volume of space (in the form of 2D

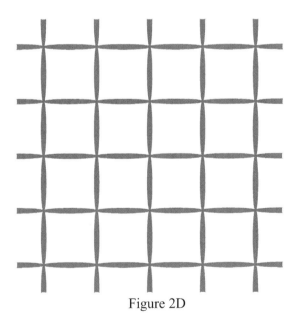

Figure 2D

energy fields) cannot express itself to a 3D being. However, a 3D matter particle of the same volume is acted upon by the 2D energy fields. 3D matter particle is able to express itself to an observer. We recognize the matter-bodies by its expression to an observer. Even though the matter content of a volumetric space in the 2D energy fields remains hidden from observers, a 3D matter particle of the same volume within the 2D energy fields is observable. This is why the 3D matter is considered as real matter and 2D and 1D matter are considered as functional matter, in this concept.

Figure 2D, shows quanta of matter in a number of junctions, together, forming part of the latticework (or network) structure of a 2D energy field in a plane. Since the constituent quanta of a stable 2D energy field latticework have almost identical magnitude of matter content, they are of equal length. They can also be regarded as forming the sides of geometrical figures of squares in the latticework. 2D energy fields extend infinitely in each of the 2D spatial dimensions in every plane in 3D space. Movements of constituent quanta of matter in the latticework of a 2D energy field or any effort, applied by them, can act only in the plane containing the 2D energy field and through the medium of the latticework of the 2D energy field in that plane. Entire space in the universe (and beyond, in cases of multi-verse) is filled with 2D energy fields and quanta of matter in this manner. 2D energy fields (and their constituent quanta of matter) in different planes, that are intersecting each other, co-exist without mutual hindrance.

Co-existence of quanta of matter in different planes at a common point may be illustrated by an example in 3D space system, as viewed by us, the 3D beings. Let there be two planks 'A' and 'B' intersecting each other at an angle. In order to do so, at least one of the planks has to be parted so that the other can pass through. Let the plank 'A' be parted into two, where plank 'B' is intersecting it. Now the plank 'A' is in two parts and there is a definite distance or gap between the two parts of plank 'A', through which plank 'B' passes. As the thickness of the plank 'B' is reduced, gap in plank 'A', required to let plank 'B' pass through, decreases. When the thickness of plank 'B' is reduced to less than the thickness of an atom, plank 'A' will (almost) have continuity, because partition of plank 'A' by plank 'B' will no more isolate atoms of plank 'A' on either side of the gap. If the thickness of plank 'B' is reduced further towards nil and the atoms of plank 'A', situated at the partition are able simultaneously to be part of materials of both the planks 'A' and 'B', it is as if the plank 'A' is not parted at all and is continuous. Although the plank 'B' is now passing through the plank 'A', which is continuous, it is not an obstacle to the planks 'A' or 'B' being continuous. In other words, at their point of intersection, both planks 'A' and 'B' co-exist. If both planks are of zero thickness, these planks can co-exist at an intersecting point without affecting each other's continuity. Likewise the 2D energy fields with common intersecting point can be considered to exist (within this restriction) simultaneously in all directions and in all planes in our 3D space system.

Entire space is filled with 2D energy fields, constituted by the quanta of matter. Attempts to determine the total mass or the matter density in any part of the space should take these factors also into consideration. Because; the quanta of matter are matter particles (in their functional state) and all matter particles have matter contents and mass, however small they may be. This part of matter, in its functional state and invisible to us (the 3D beings) constitutes the 'dark matter' in the universe. Here, magnitude of dark matter is in relation to the total 'visible' matter content existing in the universe. It is not related to the gravitational actions, required to determine any assumed state the universe.

We are part of a three-dimensional world. We are unable, by our senses, to consider anything but the 3D matter as real. A quantum of matter becomes real, in our sense, only when it starts to grow into third spatial dimension to form 3D matter. As far as our senses are concerned, quanta of matter in their one-dimensional and two-dimensional spatial states and hence all 2D energy fields may be considered as purely functional or ethereal matter than real matter. They are not virtual or imaginary. They are considered to be functional, only because of our inability to detect their physical presence in their 1D and 2D states. Our present detecting instruments and measurements are not suitable for these 1D and 2D entities. They will become suitable only when we will be able to define the thickness of a plane or breadth and thickness of a line in terms of our measuring units.

2.3.1. Reaction:

Once a 2D energy field is formed, it tends to maintain its continuity and stability. To move a quantum of matter within (or detach one quantum from) a 2D energy field requires effort. Certain work has to be done to accomplish this. Movement of quanta of matter from their current location is a work. An attempt to move a quantum of matter, which is part of a 2D energy field, is opposed by the aligning efforts, present between the quanta of matter in the latticework. Magnitude of this opposing effort is just sufficient to prevent or to restore the movement of the quantum of matter within the 2D energy field. This is the 'reactive force' or reaction applied against an external effort (for the time being, imagined as being applied directly on to the quantum of matter).

Movement of a quantum of matter within (or from) the latticework structure of a 2D energy field can take place only if the external effort overcomes the reaction to it. Rate of movements of quanta of matter in the 2D energy field latticework, in relation to space (their neighbors) in terms of rate of distance moved, is the 'force' and in terms of time, it is the 'power'. Displacements of quanta of matter, within the latticework structure of 2D energy fields are the 'work-done' (on any body or by another body).

A reaction, produced in a stable system of 2D energy field, is proportional to the stress in its latticework (destabilizing action) due to distortions present in it. This stress will cease on removal of the distortions (cause of the reaction) in the 2D energy field. This phenomenon gives rise to the axiom that *"all actions have equal and opposite reaction"*. Reaction in a region, trying to restore the equilibrium of the 2D energy field, also produces work but in opposite direction to the work developed by external effort. If the work, produced by the reaction is equal to the work, produced by external effort, system of 2D energy field remains in balance. A distorted 2D energy field, due to its inherent property to remain homogeneous and isotropic, disperses any distortion in it unless such dispersion is prevented by an external mechanism.

2.3.2. Work and force:

If an action and its reaction on a system do not balance, the system will breakdown. As soon as the action exceeds the reaction, certain permanent work is done (or stored) in the system. Permanent work remains with the system, in the form of changes in its state, until it is removed by a work-done in the opposite direction by another external action. Any such change in the system is the 'work-done' by the action of an effort. As long as an action continues, work will continue to be done. If the reaction equals a continuing action, no further work is done. Effort, to do more work, is present but it is neutralized or balanced by the reaction available. Any work-done in a system is maintained steady until it is modified by another external action. While considering displacement or motion of a body, it is prudent to neglect all 2D energy field distortions required for the body's creation and integrity. We shall consider only those additional 2D energy field distortions, which are associated with the body and required for the body's state of motion.

Work is always related to certain motion or displacement. Rate of work-done with respect to the (rate of) displacement is the 'force'. When the work is related to time, its rate is the 'power'. A force may be understood as the rate of production or release of additional distortions in the 2D energy fields, with respect to the speed of the body. [State of motion of a body (its speed) indicates the magnitude of additional work contained about the body]. It is a mathematical relation. Relations are functional entities. Force is a functional entity, which gives the rate of change in the magnitude of work-done with respect to the (rate of) distance moved by the body in unit time. Since the force is a functional entity, describing the relation between two entities, it has no real existence and it can neither act nor be acted-on physically. It exists only in mathematics and in the minds of rational beings. Force shows relation between work and (change in the) distance moved in unit time, both of which are real entities. Distance is the separation between two points in space. It is tangible and real. Work is the result of an action. Hence, a force can be understood, in general terms, as the cause of an action.

Work is the total magnitude of additional 2D energy field-distortions in association with a body that produces body's displacement or motion. It is a real entity and it can be transferred or transmitted from one body to another or from one place to another. When we say that a force is transmitted, it is the work that we are transmitting, by transferring 2D energy field-distortions (work-done) of one latticework square to its neighbors, one after the other, in the same 2D energy field. Since the force is one of the rates of work-done, it is in existence only when the work is being done. A force may be considered as 'acting' or in existence only when it causes additional distortions in any part of the 2D energy field. Otherwise, though the effort (ability to do work or stored energy) is present, the force cannot be considered active or in existence. As long as the effort is maintained, the force may be considered as being 'applied'. When a force is active, it does work and when a force is applied, it does not do work. Since no work is done during application of a force, the force is non-existent during such time. However, since the ability or effort to do the work (at the same rate as can be during its action) is present, the force is considered as present in a non-active state. In true sense, the force comes into existence only when a work is being done.

Since an external force is applied against the stabilizing force of an inherently stable system, the reaction developed is always equal and opposite of the destabilizing force. That is, the magnitude of reaction is just enough to bring back stability and remove the 2D energy field field-distortions in the system. Only when one of them exceeds the other, the force or reaction become active and certain work is done or undone. Conventionally, the work is regarded as a result of an action by a force. That is, the force or power is considered as primary entity and the cause of work. By this concept, work is the primary entity and the force or power, being the rates of work-done, is the result of a work-done. Both, the force and power are functional entities. Work, on the other hand, is physical displacements of quanta of matter in 2D energy fields. It is real. If we could see the quanta of matter and their displacements in the 2D energy fields, an observer could see or measure a work-done.

Transfer of work (additional 2D energy field-distortions) from one place to another produces apparent interactions between bodies. Force or power is the measure of rate of the transfer of work. In this text, the term 'force' is also generally used to represent (rate of) transfer of work-done (energy) from a 'force-applying body or mechanism' to a 'force-receiving body or mechanism'. Distinction between the references used to define force and power (displacement and time) is not always considered. As a result, the terms 'force' and 'power' are used synonymously. In general terms, they represent (rates of) actions of an effort. A reaction, exerted by the 2D energy field during its deformation, gives rise to reaction or 'reactive force'. It is the result of the strain set up in the quanta of matter of the 2D energy field-latticework. All natural forces in the universe are produced in this way.

2D energy fields are inherently stable systems. Distortions introduced into them cause instability. Quanta of matter, forming the distorted region of a 2D energy field, will be under stress to return to their stable configuration. If left free, a region of the distorted 2D energy field will gradually return to its stable state under the action of the strain from the quanta of matter at their junction points. Restoration of stable state will release stress in the 2D energy field. Stress experienced by the constituent quanta of matter in a distorted region of 2D energy field is the 'energy' stored in that region. Energy has no separate or independent existence. Energy is a functional entity that indicates the stress in the universal medium of 2D energy fields due a work done. Depending on the phenomenon, associated with the cause of additional distortions in the 2D energy field, energy may be classified into many types. Energy associated with a macro body is experienced (or stored in the 2D energy fields) about the macro body. Magnitude of energy about a macro body corresponds with the magnitude of 2D energy field-distortions in the region (work about a macro body) of the macro body. Hence, both work and energy may be used as synonyms. They are measured in the same units.

2.3.3. Self-sustenance of 2D energy field:

If a quantum of matter is removed from the latticework of a 2D energy field, its place in the latticework becomes vacant. Due to the break in the quanta-chain, adjacent quanta of matter on either side of the gap within the 2D energy field are now positioned at the ends of their quanta-chains in the same spatial dimension. They are under no compression from the quanta-chain from the end, near the gap and the parts of the quanta-chain are now free to grow in length, into this vacant place. This can be done only by moving other quanta of matter of the junction points, which are positioned at right angle to direction of such growth. Movements of quanta of matter at a junction point are restricted by the stability of other junctions to which they are part of. Hence, the growth in length, by these quanta of matter into the vacant place is against the reactions of the quanta of matter, which are placed at right angle to them in the latticework structure. Correspondingly, quanta of matter situated at right angles are angularly deflected to facilitate restricted displacement of junctions in the quanta-chain towards the vacant space in it.

Growth of quanta of matter, on either side of the gap, into the gap reduces the compressive pressure on other quanta of matter in the quanta-chain, placed in the same spatial dimension, letting them

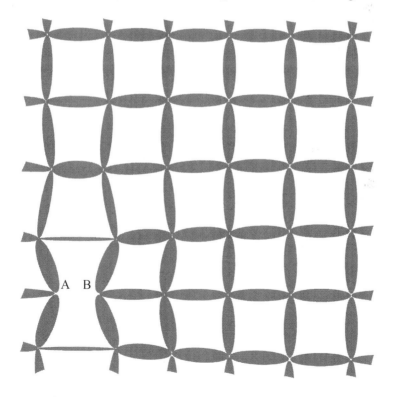

Figure 2E

also to elongate. Infinite extent of quanta-chain restricts such elongation of quanta of matter by applying reactive efforts at every junction along the length of the quanta-chain. Figure 2E shows part of a 2D energy field with a quantum of matter between junctions 'A' and 'B' missing. Probable distortions in the latticework are as shown in the figure. Displacements of the quanta of matter and changes in their width, shown in the figure, are highly exaggerated. All quanta of matter, which come under higher compression from their ends, temporarily grow into second spatial dimension. All those quanta of matter, on which the compressions from the chain are reduced, shrink in their second spatial dimension and further reduce their 'negligible' width, while growing in their first spatial dimension. Widths of the lines in the figure represent the widths of quanta of matter, corresponding to the compressive pressure on them.

Gap, formed in the 2D energy field, due to the missing quantum of matter between junctions 'A' and 'B', introduces certain stress into all quanta of matter in the neighborhood, until one quantum of matter from the chain in the same spatial dimension, slips into the vacant place to restore continuity of the 2D energy field's structure. Now another place in the latticework has become vacant. The above-described process will repeat along the quanta-chains, indefinitely, until the continuity of the 2D energy field can be fully restored. This process is completed when a free quantum of matter or a corresponding gap in the 2D energy field is found somewhere along the direction of transfer of quantum of matter's motion.

When a free quantum of matter is introduced into a 2D energy field latticework, in its spatial dimension, it produces stress in the 2D energy field until it is absorbed and made a part of the 2D energy field. Since there are no free quanta available in space, such incident can take place only when a quantum of matter from a 3D matter-body detaches from the body and finds itself free in a 2D energy field. This quantum of matter, being a part of the 3D matter-body, at the instant of its liberation, is itself a 3D object. Its magnitude, in the first spatial dimension (length), is extremely small compared to other quanta of matter in the 2D energy fields. Therefore, there is no possibility of a quantum of matter longer than the diagonal of a latticework square being introduced into a 2D energy field. A free quantum of matter in space starts its interaction with the first quantum of matter it encounters (due to its growth in length), in its own 2D spatial dimensional system. From then onwards, all its interactions and movements are limited within the 2D plane, where it has established contact with the constituent quantum of matter of a 2D energy field.

A free quantum of matter in a 2D energy field plane immediately starts to grow in its first spatial dimension as its measurements in other spatial dimensions reduce and disappear. This quantum of matter continues to grow in length until both of its ends meet quanta of matter in one of the 2D energy field latticework. First contact, by any one of its ends, determines the plane (2D energy field) for all its future actions. Once the meeting takes place, ends of the free quantum are apparently attracted to the nearest junction point. Since this quantum of matter cannot attach itself to a junction and be parallel to another quantum of matter, it will find itself in diagonal position within one of the 2D energy field latticework square. Each of its ends has joined a junction point. Both of these diagonal junction points (diagonally across the square) now have an additional quantum of matter each.

Presence of an additional quantum of matter at the junction point, in a 2D energy field, produces stress in the latticework of the 2D energy field and ultimately causes any one of the junction points to breakdown. One of the original quanta of matter, at the junction, breaks away from the junction and moves to accommodate the newcomer in its place. Only an angular movement of the quantum of matter is required for this to happen. The quantum of matter, which has broken away from the junction, now becomes a free quantum of matter, to repeat the same process of migration to the next junction point. These actions continue sequentially until continuity and stability of the 2D energy field is restored. Effort, required for this migratory movement, is derived from the motion of first quantum of matter, which joined the latticework of the 2D energy field, in free state.

Formation of stable 2D energy fields and the junctions in them are also possible with six quanta of matter making up every junction and a latticework formed by equilateral triangles of quanta of matter. 2D energy field or part of it, formed thus, is rigid and the transmission of work through them is nearly impossible. Such parts of 2D energy fields, being rigid, tend to prevent transmission or motion of 2D energy field's quanta of matter through them. Random passage of a 3D matter particle through this rigid part, breaks it down to free the constituent quanta of matter, which may then regroup and form part of the 2D energy field squares. Nature has plenty of moving 3D matter particles to accomplice this in a short time. However, due to slow and gradual actions during the formation or restoration of broken-down part of a 2D energy field, it is more probable for the quanta of matter to form 2D energy field latticework with quanta of matter arranged as the sides of squares rather than triangles. Part of a 2D energy field with higher number of quanta of matter at its junctions will gradually convert itself into more stable form of 2D energy field with junctions having four quanta of matter, each. 2D energy fields are self-sustaining entities. They tend to maintain their homogeneity and isotropy.

2.3.4. Field force:

Junctions of a 2D energy field latticework, formed with more or less than four quanta of matter to each junction and the junctions with angular difference differing from 90° between adjacent quanta of matter are unstable. They introduce additional stress in all participating quanta of matter at the unstable junctions. Reactions due to this stress are such as to restore the relative positions of the quanta of matter at the junctions to their stable and natural arrangement. In stable state, each junction will have four quanta of matter and adjacent quanta of matter at every junction will be perpendicular to each other. Reactions, produced by the distorted parts of 2D energy fields are the basis of all actions (forces) in nature. Since these 'forces' are developed within and by the 2D energy fields, they may be called 'field forces'. Currently, field forces in different perspectives are understood in various forms and they are considered as separate types of forces in nature. They are called as gravitational, electric, magnetic, nuclear, inertial, mechanical, potential, kinetic, etc. forces. These forces are synonymous and are interchangeable in most cases. It is the nature of (causes of) their actions, which differentiates them into different types of natural forces, as understood presently. Since the mechanism and actions of all kinds of forces are similar, fundamentally, there is only one type of natural force.

We recognize actions by movements of 3D matter-bodies. Since transfer of distortions in 2D energy fields moves matter-bodies, motion of matter-bodies bear all properties associated with the nature of transfer of distortions in 2D energy fields. Nature of transfer of distortions in 2D energy fields is inertial. Hence, all movements of matter-bodies in nature are inertial. Whatever causes movements of a 3D matter-body, we recognize the action only when a matter-body attains inertial motion. Motion at a constant speed (rate of displacement) is provided by the property of inertia. Force is the change in the rate of displacement of a matter-body in relation to the 3D matter-body's matter content. That is, whatever be the cause, force is nothing but the relation between acceleration of a 3D matter-body and its matter content. Causes could be differentiated into gravitational, electromagnetic, nuclear, inertial, etc. actions, but the force (being a mathematical relation) remains the same.

2.3.5. Stabilization of 2D energy field:

For easier explanation, nature of quanta of matter and their interactions may be considered as follows. Each quantum of matter may be considered as a one-dimensional body with two ends. As part of a 2D energy field, its body can accommodate elongation under tension. When under compression from the ends, its length may be reduced, by its body growing into the second spatial dimension. Ends of different quanta of matter (apparently) attract each other and their bodies (apparently) repel each other, in the same 2D plane. Repulsion between their bodies is only apparent and is the result of attractions at their ends, which tend to align the quanta of matter in contact, to each other in the same spatial dimension. Further on in this text, properties of quanta of matter are considered to be thus. However, it should be clearly understood that in reality, quanta of matter cannot act through empty space and can attract only on those quanta of matter, which are in direct contact. Apparent repulsion between their bodies is the result of apparent attraction between their matter contents through their ends, when in contact.

Assuming the above given apparent properties for the quanta of matter, it can be seen that, when they form the squares of a 2D energy field latticework, the opposite sides of a square appear to repel each other and the opposite corners of a square appear to attract each other. An external effort (considered as a separate entity), acting on a quantum of matter at a junction in the 2D energy field latticework, produces its movement in relation to the other quanta of matter of the latticework. Since every quantum of matter in a 2D energy field, is connected to others through junctions and quanta-chains, movement of any one of the quanta of matter introduces movements of all neighboring quanta of matter also. Such movements produce angular differences between quanta of matter at the junctions, which in turn produce reactions due to stress in the latticework structure of the 2D energy fields.

Reactions, applied by the quanta of matter at a junction, to restore their stable state, are proportional to the angular difference between their present states and their stable states. In other words, it can be stated that the displacement of a quantum of matter in relation to its neighbors in a 2D energy field produces the 'reactive force'. Latticework-structure of a 2D energy field makes it elastic and resilient, up to a limit, able to absorb movements (strain) of any constituent quantum of matter. Displacements of one quantum of matter in the latticework are passed along to neighboring quanta of matter in the same latticework. Due to its latticework structure, a 2D energy field has the qualities of both solid and fluid bodies. Rigidity of the quanta with regard to bending or shearing movements bestows the 2D energy fields with its solid property. Fluency and compressibility of the latticework formation bestow the 2D energy field with its fluid property.

Once a junction is formed by at least two quanta of matter, further additions of quanta of matter to the junction can take place only in the two-dimensional space system, formed by the first two quanta of matter. While forming a junction, first two quanta of matter have already set the 2D space system for their future interactions with other quanta of matter. These quanta of matter can apply reaction or move themselves only in a plane in the two-dimensional space system set by the quanta of matter at the junction. Therefore, subsequent additions of quanta of matter to this junction also have to conform to the two-dimensional space system at the junction. Thus, quanta of matter cannot form 'energy fields' in any higher dimensional space system, other than the two-dimensional space system. Quanta of matter form separate 2D energy fields in every plane in the space.

Formations of 2D energy fields, mentioned above, are only hypothetical cases, because no new 2D energy field can ever be formed in nature. 2D energy fields already fill the entire space and no additional 2D energy field can ever be formed within this space. They are always there and are forever. Above description of development of a 2D energy field is speculated only to understand the local breakdowns and restorations of 2D energy fields and their own actions during stabilization. 2D energy fields have no beginning or end, in both extent and time. They are infinitely vast and have perpetual existence. Everything real, in nature, is created out of and by the 2D energy fields. 2D energy fields are the basis of all actions and interactions in nature. They create, provide for and sustain every material object in the universe. All types of energies are originated, stored, transmitted and dispersed through and by the 2D energy fields. 2D energy fields are instrumental to gradual and cyclic destruction and creation of 3D matter (macro) bodies in nature. In short, 2D energy fields provide an all-encompassing universal medium in nature.

2.3.6. Equilibrium of 2D energy field:

Stable 2D energy fields (or their parts) are formed by quanta of matter of equal matter contents. Quanta of matter, which differ from average size in a 2D energy field, produce uneven latticework squares. Such parts of 2D energy fields are unstable. If unevenness is large, the offending quantum of matter is shifted along the latticework until it, in conjunction with other similar quanta of matter, tends to create a 'disturbance' in the 2D energy field. Such a disturbance becomes focal point for probable creation of a basic 3D matter particle (photon). For a stable part of 2D energy field, to be in balance within itself, each constituent quantum should form the side of a perfect square in its latticework structure. (Hence, the name latticework squares. Though, when under strained condition they are more like parallelograms or other geometrical forms with straight sides). In stable condition of a 2D energy field, all moving tendencies (stress) on quanta in the latticework neutralize each other and the 2D energy field becomes isotropic and homogeneous. This state of equilibrium is self-sustaining and it will be the endeavor of every 2D energy field to maintain this state of serenity. This characteristic property of the 2D energy fields is the cause of all actions and apparent interactions in nature.

Figure 2F shows one square of a vast 2D energy field-latticework structure. 'Q_1', 'Q_2', 'Q_3' and 'Q_4' are four quanta of matter of equal matter content, forming one square of the latticework. A, B, C and D are the junction points to which these quanta of matter are attached. In equilibrium state, all apparent actions within the square are in balance and there is no resultant displacement in any direction. Opposite corners 'A' & 'D' and 'B' & 'C' may be considered to be attracting each other. Similarly, opposite sides of the square may be regarded as repelling each other. From the figure, it can be seen that all these apparent actions, shown by dashed arrows, being equal and opposite, neutralize each

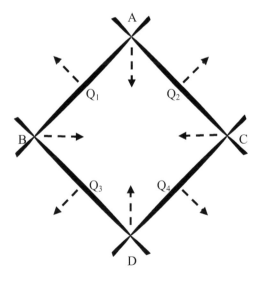

Figure 2F

other and the 2D energy field latticework square remain in stable state.

Let there be an external effort, acting on the junction of quanta of matter Q_2 and Q_4 of a latticework square, as shown in figure 2G. (All forces are applied through the medium of 2D energy field only. Direction of the force is controlled by the relative magnitudes of displacements of the quanta of matter at a junction. Here, for explanation, we are assuming an independent and hypothetical force, in the direction shown by the block arrow, acting through empty space on the latticework square). This force has components in opposition to the existing apparent repulsion on the arms AC and CD of the latticework square and thus reduces the apparent repulsion on these arms. These also may be regarded as assisting the apparent attraction between corners 'B' and 'C' of the latticework square. Let the corner 'B' be held steady. The side AC will turn about the corner 'A' and the side CD will turn about the corner 'D'. Angular movements of quanta of matter AC and CD bring the corners 'B' and 'C' nearer and take the corners 'A' and 'D' farther. Angles 'A' and 'D' decrease and the angles 'B' and 'C' increase. Changes in the divergence angles between quanta of matter of the latticework square give rise to increased strain in them and thus, produce resultant reaction. Directions of resultant of these reactions are in direct opposition to the applied motion (force) at junction point 'C' and act such a way as to restore the equilibrium of the latticework square. Arms of a deformed latticework square remain under stress as long as the latticework square remains deformed. The stress, produced by the deformation of the square, acts as its desire for the latticework square to return to its stable state.

Other junction points 'A', 'B' and 'D' (with the help of quanta of matter at junctions beyond them) restrict the mechanical deformation of the latticework square. These junction points are also shared by other quanta of matter in the latticework beyond the latticework square considered in figure 2G. Reactions from these quanta of matter restrict free motion of the junction points. Restrictions appear as axial compression or tension on the quanta of matter of the latticework square. These, in turn, pressurize respective quanta of matter to grow temporarily into second spatial dimension or reduce compression from their ends to elongate them further. Thus, part of the external effort is used to convert the quanta of matter into second spatial dimension and the stress or energy is stored in the quanta of matter in the form of pressure energy, producing the changes of their spatial dimensional states. As and when it is possible, this pressure energy is returned by the quanta of matter, while they regain their original states.

Work, put in by the external effort, is stored in the form of reduced length (pressure energy) in the quanta of matter to change their spatial dimensional status and in the form of relative displacement of the quanta of matter within the latticework. These two changes together constitute the work-done by the external effort. Because of the latticework structure of the 2D energy field, work-done in any part of the latticework cannot remain isolated or permanent, unless it can be maintained by external means. In free space, latticework distortions tend to spread-out in the 2D energy field.

Under extreme conditions, deformations produced in a 2D energy field may be large enough, so that some of the junctions of

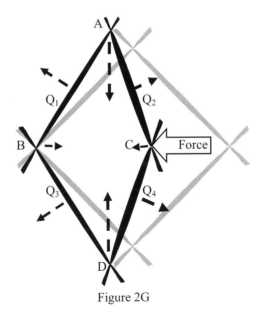

Figure 2G

quanta of matter in the latticework may be compelled to accommodate more number of quanta of matter or abandon available quanta of matter to form additional rectangles or triangles in the latticework to prevent its breakdown. Such additional formations remain under stress and their equilibrium can be restored only when the strain caused by such formations and the distortions due to them are removed from the latticework. If the source of deformation is too great, latticework squares of the 2D energy field break down (locally) and release constituent quanta of matter to be free in the space. Magnitude of the local breakdown in a 2D energy field depends on the 'power' of the work input.

Consider a latticework square, as shown in figure 2G. Let the junction point 'B' is prevented from moving and the junction point 'C' is displaced towards 'B'. The external efforts transmitted through the quanta of matter Q_2 and Q_4 cause outward displacement of junction points 'A' and 'D'. Magnitude of displacement of each of these junction points is proportional to the cosine of angle between the line joining points 'B' & 'C' and the respective quantum of matter. If the junction points 'A' and 'D' of the latticework square are also prevented from moving, junction point

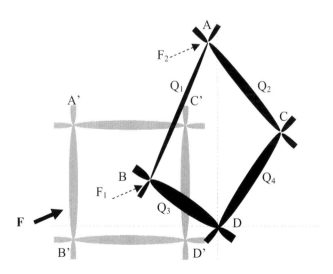

Figure 2H

'B' imitates displacement of point 'C' (the latticework square behaves like a solid body). If junction points 'A', 'B' and 'D' are prevented from moving, attempt to displace junction point 'C' to the left produce compression of all the quanta of matter in the latticework square. Behavior of the latticework square, in this case, is exactly as a fluid body behaves under an external effort. Hence, 2D energy field, though it is made up of seemingly solid and rigid matter particles, behaves like a fluid body. In reality, only 3D macro bodies can exist in various physical states. In case of 2D energy fields, solid and fluid states are mentioned to describe their properties rather than their physical states.

Figure 2H shows the representation of probable distortion of latticework square A'B'D'C' in a 2D energy field under the action of an external force, F. Components of external effort, F_1, F_2, etc. acts at all junction points in the region. Depending on the relative locations, magnitudes of these component efforts will vary. Under the action of the external effort, junction points are displaced until reaction from the latticework equalizes the applied external effort. Probable transformation of the latticework square A'B'D'C' is as shown by ABDC in the figure.

2.3.7. Properties of 2D energy fields:

A 2D energy field is a two-dimensional entity. It has only length and breadth as its fundamental spatial dimensions. A real entity in space essentially exists in all spatial dimensions. Hence, however small the dimensional measurements may be, a 2D energy field has its existence in the third spatial dimension also. A volumetric space is made up of great many parallel planes, in contact. If the thickness of a plane is considered as nil or zero, any number of parallel planes cannot constitute a volumetric space. However, 2D energy fields in space constitute volumetric space. Therefore, a 2D energy field or other 2D matter-bodies do have certain thickness, negligible with respect to 3D measuring systems.

2D energy fields have the following inherent properties:

1. Inherent properties of 2D energy fields are derived from inherent properties of their constituent quanta of matter and their mechanical structure in the form of latticework.

2. 2D energy fields are two-dimensional entities made up of single-dimensional quanta of matter. Each 2D energy field exists and acts in its own plane.

3. They extent infinitely in all directions, each one in its own plane and tend to remain homogeneous and isotropic. They are infinitely vast and have perpetual existence. 2D energy fields fill the entire space. No 3D matter particles can exist outside 2D energy fields. On the whole, 2D energy fields are perpetual and steady in space. No new 2D energy field is ever produced. They provide an absolute reference.

4. Only one 2D energy field exists in any one plane and all planes in all directions in 3D space contain one 2D energy field each.

5. 2D energy fields in different planes, passing through the same point in space, co-exist at the point.

6. Quanta of matter, in a 2D energy field, are held under compression from their ends. Quanta-chains, in the same plane, cross at junction points in perpendicular directions.

7. 2D energy fields are self-sustaining entities. They strive to sustain their integrity, stability, homogeneity, isotropy and serenity. Each 2D energy field has apparent adhesive nature within itself and tends to maintain its continuity in the plane of its existence.

8. In stable state of a 2D energy field, constituent quanta of matter form sides of perfect squares in the latticework structure. A change from the stable state produces restoring reactions in the latticework structure.

9. Interactions between two points in a 2D energy field are confined to the plane containing both the points. [In order to avoid theoretical possibility of more than one 2D energy field passing through two points, here, a point should be understood to be having the smallest area and be a part of one 2D energy field plane. So that, there can be only one 2D energy field passing through any two coplanar points.]

10. 2D energy fields tend to reduce any disturbances within themselves to a minimum. This is achieved by forcing a disturbance to reduce its size to minimum and (when possible) expelling this area of disturbance, from the 2D energy field. Tendency of 2D energy fields, to close-in any gap in them, produces gravitation. 2D energy fields tend to reduce disturbances in them to minimum either by reducing the sizes of the disturbances by shaping them circular and compressing to smaller size or by ejecting the disturbances out of themselves.

11. All higher dimensional space systems exist within the 2D energy fields and all bodies in higher dimensional space systems are 'disturbances' with respect to the 2D energy fields.

12. All 3D matter particles are created from, sustained by and reverted back into the 2D energy fields.

13. 2D energy fields provide an all-encompassing universal medium for all apparent interactions between 3D matter-bodies.

14. Region of 2D energy fields, about a 3D matter-body, store work-done about the body in the form of distortions (energy in the form of stress due to the distortions) to sustain the integrity and stability of a matter-body and its state (of motion).

15. Distortions (causes of field forces by their interactions) in two 2D energy fields cannot interact. Transfer of distortions or interactions between distortion fields are limited to own plane of each 2D energy field. Simultaneous actions in many planes appear to be an action in 3D space system.

16. 3D matter particles are displaced in space by the transfer of distortions in steady 2D energy fields. Absolute motions of matter-bodies are with respect to the steady (parts of) 2D energy fields. 3D matter-bodies are moved by the 2D energy fields rather than the bodies move through 2D energy fields.

17. Latticework structure of a 2D energy field causes sequential development of distortions in neighboring latticework squares. Distortions, once developed, remain permanently within the 2D energy field and continue to transferred at a constant linear speed, unless removed by external action. These phenomena give rise to the property of inertia.

Since 2D energy fields fill the entire space, the term '2D energy fields' can be used as synonymous to 'space'. They mean the same thing or they indicate the same region. Space is a non-entity, presupposed by rational beings, whenever a 3D matter-body is envisaged. Unlike space or aether, the 2D energy fields not only provide for a region of existence for the 3D matter-bodies but they also have definite structure and properties to account for all apparent interactions between 3D matter-bodies. By this, the space or aether (undefined terms in the past) attains definite structure and properties. Having a structure and real existence, the 2D energy fields (unlike space) can be deformed. They can bend, curve, contract or expand, as is required in some of theories involving the undefined space. 2D energy fields constitute an all-encompassing universal medium. Thus, they substitute for the undefined 'aether' in present aether-theories and for intermediary in present 'field theories'.

2.4. Production of a Disturbance:

Quantity of matter, contained in a quantum of matter, may vary from one quantum of matter to another. Since no new quanta of matter are ever created, we can say that the quanta of matter exist eternally in the universe. There is no mechanism to regulate the matter content of a quantum of matter during their origin or later. Generally, majority of the quanta of matter are identical in their matter contents and tend to form stable 2D energy fields. However, there are many quanta of matter, whose matter content varies from the average. Presence of quanta of matter of different matter contents in the latticework structure distorts a part of 2D energy field. These quanta of matter, when part of the latticework, produce strain in the 2D energy fields. This part of the 2D energy field remains under stress. As and when an opportunity arises, strained parts of a 2D energy field join together to cause local breakdown in that 2D energy field.

Presence of a quantum of matter, with higher matter content in it, causes strain in a 2D energy field. Due to the strained state of a 2D energy field, a quantum of matter with higher matter content has higher inward pressure at its ends. Higher inward pressure at its ends causes greater growth of the quantum of matter into its second spatial dimension. Existence of certain quanta of matter in a 2D energy field in higher-dimensional state causes interference between neighboring 2D energy fields (through its matter-body) and may cause their local breakdown.

Quanta of matter at the junction points in a 2D energy field are held together by the feeble apparent attraction between matter contents of the quanta of matter. There is no rigid bond between them. Speed of their relative movements and their bonding forces are limited. Excessive force/pressure than that can be born through a 2D energy field may also cause its local breakdown.

External actions on a part of the 2D energy field may disturb its stability and structure. Deformation in the latticework structure introduces strain into the 2D energy field. If the stress produced by such a strain is strong enough, the 2D energy field in that area may locally breakdown.

Local breakdown of 2D energy fields is the first step towards creation of 3D matter particles. During any such breakdown, 2D energy field in that region disintegrates and constituent quanta of matter are liberated from its latticework structure. These detached quanta of matter become relatively free and float around in the space, before they can regain their place in the latticework. A group of quanta of matter, within a gap created by a local break down in a 2D energy field, forms a 'disturbance'. These quanta of matter may exist in their free states or they may exist as a single body in the form of higher-dimensional matter particle. Development of a disturbance initiates the creation of basic 3D matter particle. A 2D disturbance breaks the continuity of 2D energy field of the plane. A 3D disturbance simultaneously breaks the continuity of all the 2D energy fields, in all the planes passing through it.

Should there be too many loose quanta of matter, in a place (pocket or gap formed by a local breakdown) in a 2D energy field at the same time; process of their assimilation into the 2D energy field is too slow for all of them to be assimilated into 2D energy fields. These free quanta of matter now crowd in the pocket within the 2D energy field to form a 'disturbance' within the 2D energy field. A 'disturbance' in a 2D energy field is a group of quanta of matter, which affects 2D energy field's stability and serenity. Due to the tendency of quanta of matter to grow in their first-dimensions, each one of the free quanta of matter, within the gap, starts to grow in its own dimensional space system. Free quanta of matter in the plane, together, exert an effort on the 2D energy field latticework of the plane to expand the pocket of their existence.

Since the quanta of matter in the disturbance and the 2D energy field are in the same plane, they cannot co-exist. 2D energy field, by its inherent nature, now exerts a reaction onto the disturbance from all around, to contain it. Because of the discontinuity, the 2D energy field all around the gap tends to extend into the gap to re-establish its continuity. This action is against the free quanta of matter's attempt to enlarge the gap. Intrusion of the 2D energy field into the gap reduces the size of the gap and the disturbance in it. This is the basis for the inherent property of the 2D energy fields to reduce magnitude of disturbances in them, to a minimum.

Although it is the nature of the 2D energy fields to remain calm and serene, disturbances may occur in them, spontaneously or intentionally. Sudden supply of large quantity of free quanta of matter discarded from 3D matter particles into a region of 2D energy fields or sudden absorption of large quantity of quanta of matter from the latticework cause disturbances. Presence of quanta of matter, having very large or very small matter contents, in a region of 2D energy field also produces disturbances.

Quanta of matter with very large matter contents cannot be accommodated in a 2D energy field latticework. It invades the space occupied by many 2D energy fields and creates discontinuity in all of them. Consequently, all these 2D energy fields act on the quantum of matter to compress and maintain its state in higher dimensional space system, as a disturbance. 3D matter particles are produced from the disturbances. This phenomenon assures the presence of 3D matter in nature at all times. Whatever be the initial conditions of the universe, if the is any, we can say that certain 3D matter, in whatever form, is and was always present in the nature. This also makes sure that 3D matter cannot be entirely destroyed, in nature.

2.4.1. Development of a disturbance:

Since the quanta of matter, detached from 2D energy fields (or from 3D matter particles) are free, each of them grows in length until they encounter another quantum of matter, either free or part of a 2D energy field. A free quantum of matter, making contact with another quantum of matter in a 2D energy field starts the process, for it to become a part of the 2D energy field's latticework structure. However, if the gap produced in the 2D energy field is large enough and number of the quanta of matter detached are many, developing and restoring that part of the 2D energy field takes some time. In the meantime, the 2D energy field from all around the gap tends to close in on the gap to restore its continuity.

Detached quanta of matter, within the gap in a 2D energy field, are unorganized and behave like independent matter particles, each one trying to elongate in its single-dimensional spatial system. Such a group or collection of quanta of matter is a 'disturbance'. Nature of a disturbance, formed by this group of quanta of matter, is neither homogeneous nor isotropic. Although the disturbance is also a collection of quanta of matter, it is very distinct and unorganized compared to the 2D energy field, which is an orderly collection of the quanta of matter. A collection of quanta of matter, even if it is a single body in higher-dimensional system, is a disturbance with respect to 2D energy fields. A disturbance of higher-dimensional space system simultaneously exists in more than one 2D energy field. Part of the disturbance in each plane is acted upon by corresponding 2D energy field in that plane. Since a disturbance (or part of a disturbance), in a plane, is not a part of the 2D energy field, it breaks the continuity of the 2D energy field in that plane. Hence, any object that creates a discontinuity in the structure of a 2D energy field also can be regarded as a disturbance with respect to the 2D energy fields.

It is not necessary for the detached and free quanta of matter to continue their interactions with the same 2D energy field from which they are detached. Depending upon the direction of the free quantum of matter during its expansion in the 1D space system, the first (quantum of matter of) 2D energy field, it meets is the destination for all its future interactions. A quantum of matter, finding itself at an angle to the 2D energy field from which it was detached, can interact only with a 2D energy field in the plane containing the quantum's present position.

2.4.2. Magnitude of a disturbance:

A disturbance is a collection of free quanta or a 2D particle, constituted by more than one quantum, (or part of a 3D particle in the plane) within a gap in a 2D energy field. Since the thickness of a 2D energy field is negligible, a disturbance formed in it (or part of a higher dimensional disturbance in its plane) also is a two-dimensional entity. 2D energy field around the disturbance is in contact with it all around its periphery. All interactions between a disturbance and the 2D energy field take place at the place of their (points of) contact. Magnitude of such interactions depends on the magnitude of their direct contact. A 2D disturbance is a 2D entity and a 3D disturbance is a 3D entity. Therefore, the contact between a disturbance and the 2D energy field of its existence is limited to the length of perimeter of the disturbance in the plane of 2D energy field. Thus, the perimeter of a disturbance, in a plane can be taken as its magnitude in the 2D energy field of the plane, in terms of (apparent) interactions with other entities.

Normally, all stable disturbances are circular in shape with uniform matter-density. Hence, the perimeter of a disturbance has a definite relation to its size. Size of the disturbance, together with its matter density determines the total matter content of the disturbance. Matter content of a disturbance is its magnitude in terms of matter, it contains. Since we have no dimensional measurement system, to directly measure matter content of an entity, we have to depend on indirect measurements. In this case, we can take it that the length of the perimeter of the disturbance in a plane represents the magnitude of its interactions due to its matter content and hence the magnitude of the disturbance in that plane.

This value may be modified by a constant of proportion, determined from practical observation of apparent interactions in nature, to devise a practical measurement system.

2D energy fields fill the entire space outside 3D matter particles. There is no empty space. Matter densities of a quantum of matter and that of a 3D matter particle are identical. Hence, entire space is filled with matter and matter densities in all places are identical, irrespective of the type of matter-bodies in that region. However, as 3D beings, we are able to observe only 3D matter particles/bodies. This restricts our ability to observe matter, distributed outside 3D matter-bodies.

A 3D macro body has millions of basic 3D matter particles in it. These matter particles are situated far from each other within the macro body. Because of the gaps between basic matter particles, 3D matter density of the macro body is very little compared to the matter density of basic matter particle. Matter contents of constituent quanta of matter in 2D energy fields, outside the 3D matter particles are ignored. Matter densities of all basic 3D matter particles are same as the matter density of quantum of matter. Matter density of a 3D macro body depends on many other factors including; matter contents of basic 3D matter particles, their numbers and distribution, nature of 2D energy fields around the macro body, etc. A cross sectional plane of a 3D macro body contain very little 3D matter with respect to its perimeter as compared to the matter content of a hypothetical 2D body of same size. Therefore, while considering a 3D macro body as a single 3D disturbance, its matter density and total matter content (in each plane) is much less than a hypothetical 2D body of same size.

*** *** ** *

Chapter Three
GRAVITATION IN 2D SPACE

3.1. Gravitation:

A collection of independent quanta of matter (in any dimensional space system) within a gap in a 2D energy field forms a 'disturbance'. Presence of a disturbance (or gap) creates a discontinuity in the 2D energy field of the plane. A gap in a 2D energy field (its discontinuity) offsets its stability in that region. Stabilizing actions on the squares of the 2D energy field latticework, from the direction of the gap is not present any more. In order to restore its stability and continuity, the 2D energy field exerts itself to close-in on the gap. The 2D energy field closes-in by axial displacements of all chains of quanta of matter (by self-extension of quanta of matter in them) around the gap, towards the centre of the gap. Presence of the disturbance prevents ingress of the 2D energy field into the space occupied by it in the gap. Effort by the 2D energy field to close-in continues as long as the gap exists in it, even when it is occupied by a disturbance. Tendency of 2D energy fields to close-in towards a gap in them causes the phenomenon of 'gravitation'.

Action of the 2D energy field, closing-in to fill up the gap, applies an (translational) effort or pressure on a disturbance, present within the gap, even if it is in the form of higher-dimensional object. If we are considering the action along a straight line towards the disturbance, it constitutes a 'gravitational force' (effort). If all the forces around the disturbance are considered simultaneously, they will constitute a pressure. This pressure, applied by a 2D energy field, on any disturbance within itself, is the 'gravitational pressure'. This phenomenon, producing the gravitational force or pressure, is the 'gravitation'. Gravitation is developed from the inherent desire of 2D energy fields to attain serenity.

Gravitational pressure is produced due to the inherent property of the 2D energy field to remain stable, continuous, homogeneous and isotropic. Because of this property, a 2D energy field always tends to fill up any gap in it, even if other objects occupy the gap. A 2D energy field continues to apply gravitational pressure on a disturbance until its own continuity can be restored. Its continuity cannot be restored as long as the disturbance is in existence within the plane. To invoke a gravitational action, it is necessary to have a discontinuity (a gap) in the 2D energy fields.

3.1.1. Range of gravitation:

Requirement of a discontinuity in the 2D energy field is the factor distinguishing the gravitational action from actions of other field forces (electric, magnetic and nuclear forces). Being a field force (an effort created within the 2D energy field, by itself), the gravitational action is also produced by an imbalance in the 2D energy field latticework structure. Gravitational action is one of the aspects of the field force, produced by an imbalance in the structure of a 2D energy field. Unlike the other forms of field forces (which are produced by interactions between distortion fields in a small region of the 2D energy field and have limited range in the space), entire fabric of the 2D energy field in the direction, away from the gap, applies the gravitational effort or pressure onto a disturbance in the gap. A 2D energy field extends infinitely in all directions in its plane. However far two disturbances may be, a 2D energy field passing through both the disturbances is continuous but for the gaps formed by disturbances. Distance between disturbances does not limit the effects of this 2D energy field on the disturbances. Hence, the gravitational pressure or effort is of long-range. Its range is limited only by the extent of space, which is infinite.

Gravitational action is the product of the 2D energy field. Hence, the magnitude of gravitational action is directly related to the extent of the 2D energy field in the direction of action of the force. Larger the extent of 2D energy field, greater is the magnitude of the gravitational force. In free space, the extent of 2D energy field is infinite. In other cases, the extent of 2D energy field from one disturbance to another is the distance between them. Thus, distance between two disturbances (or 3D matter-bodies) is one of the factors determining the magnitudes of gravitational effects on them.

3.1.2. Nature of gravitation:

Gravitational pressure (on a surface/perimeter) or force (in a straight line) is an inherent nature of the 2D energy field. Wherever there is a discontinuity in the fabric of a 2D energy field, 2D energy field all around the gap exerts itself on any entity that happens to be within the gap. Effort, applied by the 2D energy field, due to gravitation is proportional to the extent of the 2D energy field in the direction of the incoming effort. 2D energy field applies the gravitational effort such as to move a disturbance (the entity within the gap), away from the part of the 2D energy field, which is applying the effort. Therefore, basically, the gravitational action is apparently of 'repulsive nature'. It produces a 'push' action on the disturbance. Push gravitational action is an inherent property of the 2D energy field.

Other than the extent of 2D energy field, magnitude of gravitational effort is related to the compression of quanta of matter in quanta-chains. For any state of universe, numbers of quanta of matter in any quanta-chain remains more or less steady. Hence, the gravitational effects in the universe exhibit steady relations to parameters of disturbances. Changing the nature of 2D energy fields in the universe may vary the compression states of the quanta of matter in the quanta-chains to alter relative magnitudes of gravitational effects. Presence of large concentrations of 3D matter in the form of very large macro bodies may alter nature of 2D energy fields near these bodies. Due to different natures of 2D energy fields in different regions of space (due to presence or absence of 3D matter macro bodies) magnitudes of gravitational actions may be different in these regions.

Gravitation is not a mysterious property of mass or matter content of a body, as is currently believed. It does not emanate from the matter content or the mass of a body but from the surrounding 2D energy fields, in which the body exists. It is the result of inherent property of 2D energy fields to stabilize themselves. Hence, it can be said that gravitation is a property of 2D energy fields (the space). Existence of matter in its three-dimensional state that causes discontinuity of the 2D energy fields is the only reason for the 2D energy fields to cause gravitational effects on them. Gravitational effects are applied by the 2D energy fields directly onto 3D matter particles to which they are in direct contact. No additional carriers are required to transfer gravitational effects on or between macro bodies. There are no direct gravitational actions between different 3D matter-bodies. Each 3D matter-body is gravitationally acted upon, separately, by the 2D energy fields. Simultaneous gravitational actions on different 3D matter-bodies in space may be interpreted as apparent interaction between them.

Because of the 2D energy field's infinite extent in the space, extent of a 2D energy field, between any two 3D objects (disturbances), is always less than the extent of the 2D energy field outside these objects. Therefore, the gravitational action on each of these objects from the space between them is less than the gravitational action on them from their outer sides. This turns the repulsive nature of gravitation to appear to be an attraction between these two disturbances (objects). Two disturbances in the same plane, when pushed towards each other, appear to attract each other. Apparent gravitational attraction between matter-bodies is one of the minor aspects of the gravitational action.

Ignoring the possibility of a universal medium, when the gravitational action is considered as an attraction between two objects, it appears to originate from the participating 3D objects. This has led to the misconception that *"every body in the universe attracts every other body due to gravitation"*. In fact, it is the 2D energy fields (on the outer sides of these bodies), which are pushing them towards each other against the smaller push forces applied from in-between. There is no transfer of force, energy or imaginary particles taking place between the 3D bodies under (apparent) attraction due to gravitation. Gravitational actions take place between each of the 3D bodies and the 2D energy fields, separately. This is the reason why the gravitation appears to be acting through any screening (macro) body, however dense it may be and is able to make instantaneous changes in the magnitude of its action (force), on any modification to the parameters of one or both of the 3D bodies.

Since a 2D energy field extends in all directions and the

gravitational action is from all directions in the plane of the 2D energy field, it constitutes a pressure. An asymmetry in the shape of a disturbance can produce an imbalance in the translational efforts around it and compel the disturbance to assume circular shape. [Macro-bodies are union of multiple numbers of basic and fundamental 3D matter particles. Each of these basic 3D matter particles is a disturbance with respect to the 2D energy field. Gravitational pressure is applied on every basic 3D matter particle of a macro body rather than on the composite macro body. Hence, asymmetry in the shape of macro-bodies do not produce imbalance in the gravitational pressure on them]. Difference in the extent of the 2D energy field in any direction also produces an imbalance of the gravitational efforts. Such an imbalance tends to produce linear movement of a disturbance.

3.1.3. Strength of gravitation:

Gravitational action emanates from every point (or unit area) of a 2D energy field. 2D energy field has negligible thickness and so the effects of gravitational pressure or force are confined to the plane of the 2D energy field. All factors (including the shape), related to the contact between a disturbance and the 2D energy field, together determine the magnitude of the gravitational action on a disturbance, in any direction. Magnitude of the gravitational action, from any direction, on a disturbance is related to the extent of the 2D energy field (in that direction) acting on it, curvature of disturbance's perimeter and the angle subtended by the disturbance in the 2D energy field plane. Magnitude of apparent interactions, between two disturbances in a 2D energy field (due to the differences in the gravitational actions on them), increase as distance between them is reduced. Interaction between these disturbances has only an appearance. Actual interactions are taking place between the 2D energy field and each of the disturbances separately.

Gravitational actions (force) are enormously stronger than all other natural actions (forces). However, the apparent attraction due to gravitation between macro bodies, which is a very feeble by-product of gravitational actions, is the only gravitational action, recognized presently. It is thus that the gravitational force came to be regarded as the weakest of all natural forces. Actually, there is no attraction between matter-bodies due to gravitational or any other effects. Apparent attraction between matter-bodies is a result of simultaneous and separate translational motion of each matter-body due to difference between gravitational (or other natural) actions on either sides of each of the matter-bodies. Hence, the apparent action by gravitational efforts, which we regard as attraction between matter-bodies, is only a very minute fraction of actual gravitational action on the matter-bodies.

Application of gravitational effort is a continuous process. As long as a disturbance is in existence, the gravitational effort is effective on its periphery. Variations in the magnitude or shape of the disturbance affect the magnitude of gravitational effort, instantly. Thus, the apparent attraction between two bodies due to gravitation is modified instantaneously, on modification of parameters of either one or both of the bodies, without help from any virtual or assumed particles. Action or effect does not reach out from one matter-body to another. It is neither transmitted through empty space nor it requires a carrier particle nor a medium of transmission. Gravitational actions on each matter-body by the 2D energy fields is separate and its magnitude is affected by another matter-body only so much as it may change the extent of 2D energy fields in the direction towards that matter-body.

3.1.4. Gravitation on a point-disturbance:

A point in a 2D energy field is considered as a matter-body with an area of minimum or negligible dimensions and which is a part of the 2D energy field plane. This point may be considered as a hypothetical 2D disturbance in a 2D energy field and having minimum or negligible area. Extent of a 2D energy field in free space is infinite in any direction, in its plane. Space is considered free, when it is devoid of all disturbances (including 2D or 3D matter particles, transmitted distortions or macro bodies) other than the disturbance or macro body, under consideration. Such a region of space is filled with undistorted 2D energy fields in all directions and in all planes.

For the present discussion, only one plane and the 2D energy field in that plane are considered. Since the disturbance considered is of two-dimensional nature, other 2D energy fields, passing through the disturbance in various directions, need not be considered. 2D energy fields in other directions/planes co-exist with the disturbance (passing through the disturbance) without affecting it. Since the 2D disturbance creates no discontinuity in the 2D energy fields in other planes, there is no interaction between them. If the disturbance is of three-dimensional nature, every 2D energy field of planes passing the 3D disturbance is interrupted. Presence of a 3D disturbance in each 2D energy field breaks the continuity of its latticework structure. All of them act on the disturbance, separately, each one in its own plane. Total gravitational effect on a 3D disturbance is the resultant of actions by the 2D energy fields in all the planes passing through the 3D disturbance.

When a point-disturbance (in a 2D plane in free space) is considered, extent of 2D energy field in every direction in the plane is infinite. Gravitational action (force) is proportional to the extent of 2D energy field. Hence, the gravitational pressure acting on a point in free space is of maximum possible value and its magnitude is constant. Its magnitude depends only on the nature of 2D energy field, which may vary between different regions of space or between different periods of time. Let the value of this constant be G_1.

In figure 3A, let 'A' be a point-disturbance in a 2D energy field and let 'B' be a point of unit measure (of length) on the circumference of a circle. Circle OAB, whose center is at A and

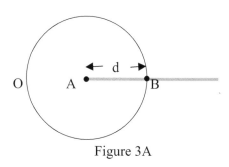

Figure 3A

has its radius AB equal to 'd'. Dimension of point 'B', being along the perimeter of the circle OAB in a 2D plane, is of single dimension - length. (An infinitesimal part of perimeter of a circle can be considered as a straight line). Gravitational force, applied by the 2D energy field, on point disturbance 'A' through point 'B' is a fraction of the total gravitational pressure applied on point 'A' by the 2D energy field, through the entire the length of the perimeter of the circle OAB.

Gravitational pressure on point A = a maximum constant = G_1

Perimeter of circle OAB $= 2\pi\, d$

Gravitational force on point 'A' through the point 'B'

$$= G_1 / 2\pi\, d \qquad (3/1)$$

Gravitational effort, acting on point 'A' through point 'B' in a 2D energy field plane, is inversely proportional to the distance between the point of disturbance and the reference point. A smaller disturbance has shorter distance from its center to its perimeter. Length of perimeter is the magnitude of a disturbance. Smaller disturbance has smaller perimeter. Gravitational pressure in free space, being constant, smaller perimeter receives greater force. Hence, smaller a disturbance is, greater will be the gravitational force felt at every point on its circumference. As the disturbance becomes smaller, gravitational force on its periphery increases in inverse proportion to its size.

3.2. Gravitation on a 2D disturbance:

A 2D energy field, by its inherent nature, acts gravitationally (applies gravitational efforts) all around a disturbance, in its plane. These efforts are directed away from the part of the 2D energy field, applying the effort. They are directed into the disturbance (towards the centre of curvature of the perimeter) from all directions, within the angle subtended by the part of the 2D energy field. Gravitational action, along any line, is such as to move the disturbance away from the extent of 2D energy field, applying the effort. Application of the gravitational effort, by a 2D energy field, is limited to its plane.

3.2.1. Shaping up a disturbance:

In a 2D disturbance, as shown in figure 3B, let AA_1, BB_1 and CC_1 be three chords of equal lengths, cutting the perimeter of the disturbance at different locations. Magnitude of gravitational action at any of these segments of perimeter, is proportional to the exposure, it has to the 2D energy field. These are in turn, proportional to the angle subtended by them to the 2D energy field, shown by the double-headed curved arrows in dashed lines. It can be seen that the segment of 2D energy field, in contact with perimeter segment BB_1, is wider than the segment of 2D energy field in contact with perimeter segment CC_1 and narrower than the segment of 2D energy field in contact with perimeter segment AA_1. Gravitational action on any point in one of these perimeter segments is a fraction of the total gravitational action on the perimeter segment. Total gravitational action on a segment, is proportional to the extent of 2D energy field that is in contact with it. Extent of the 2D energy field is partly determined by the angle subtended by the chord across any part of the perimeter. Another factor that determines the magnitude of gravitational action is the curvature of the perimeter.

Gravitational action governs the inward motion of a perimeter segment of a disturbance. Magnitude of gravitational action at the perimeter section AA_1, is more than the gravitational action at the

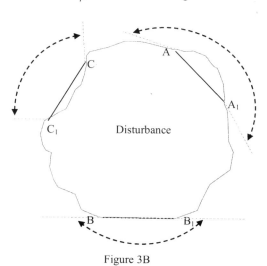

Figure 3B

perimeter section CC_1. Such differences in the magnitudes of action, over and above the compression of the disturbance by the 2D energy field, compel the disturbance to assume a circular shape. Reshaping action stops when the disturbance becomes circular and gravitational actions on all points on its perimeter become equal. Thereafter the gravitational action is only to compress the disturbance to smaller size, until all constituent quanta of matter in the disturbance are closely packed (without vacant space between them) and matter density of the disturbance equals the matter density of a quantum of matter. Once the matter density of the disturbance reaches the highest possible value (same as the matter density of a quantum of matter), gravitational pressure at the perimeter maintains the constant state of compression of the disturbance.

Gravitational action, on any part of the perimeter of a disturbance, depends also on the curvature of the perimeter segment. As it can be seen from figure 3B, a convex perimeter segment has a larger segment of the 2D energy field in contact with it and a concave perimeter segment has a smaller segment of the 2D energy field in contact with it. A larger segment of 2D energy field has greater extent and a smaller segment has lesser extent. Therefore, a convex curvature of (periphery of) the disturbance has greater magnitude of gravitational action on it than concave curvature of (periphery of) the disturbance. Higher convex curvature of the periphery of a disturbance produces greater magnitude of gravitational action and higher concave curvature causes greater reduction in the magnitude of gravitational action on that part of periphery of the disturbance.

Nature and magnitude of the curvature of perimeter of a disturbance causes an additional difference in the magnitude of gravitational action on a disturbance as explained later in the section on 'action of gravitation'. Asymmetry in the magnitudes of gravitational actions, around the perimeter of a disturbance, due to the unevenness of its perimeter during its formation, may impart linear motion to the disturbance, in its plane. This happens only when the resultant linear action is strong enough to part junction points of the 2D energy field, to pave way for the disturbance to move through it.

Similar gravitational actions, take place on a free 3D matter particle in 3D space also. 2D energy fields in every plane, passing through a 3D body, act on the body, each one in its own plane. Gravitational actions on such a body (basic 3D matter particle) try to shape the body into circular shape in each of the planes. Simultaneous actions in all planes about the body, together, tend to make a free (static) 3D matter particle (not a macro body) in 3D space, into a sphere.

3.2.2. Application of gravitation:

We may use the term 'force' to indicate an 'effort' or an 'action'. Although the gravitational action (effort) is also a field force, produced by the imbalance in a 2D energy field, it is distinct from the other forms of field forces in certain respects. Gravitational effort is a long-range force, applied continuously on 2D disturbances and on basic 3D matter particles. Every latticework square, away from a disturbance in the 2D energy fields of planes passing through the disturbance, bears certain distortions in it to cause the gravitational effort on the disturbance. Magnitude of distortions in the latticework squares diminishes progressively as the distance from the disturbance increases.

Gravitational effort is modified instantly on changing any parameter of the disturbance (all 3D matter particles are disturbances with respect to the 2D energy fields). Gravitational action by a 2D energy field is directly on to the disturbance. It requires no other medium or carrier. 2D energy fields are in direct contact with the disturbances, at every point on its perimeter (outer surface). 2D energy fields are everywhere in space and as a result no higher-dimensional matter-bodies can escape gravitational effects.

Gravitational effort is produced only when there is a discontinuity in the 2D energy field. If there happens to be a disturbance of any kind (2D, 3D or of higher dimensional matter-bodies) within the gap and it occupies the plane of a 2D energy field, the expanding quanta of matter in the 2D energy field quanta-chains are unable to pass through the disturbance to establish continuity of the 2D energy field. Expansion of quanta of matter in the quanta-chains and extension of 2D energy field into the gap is prevented by the presence of the disturbance in the gap. Effort by the expanding quanta-chains is applied on to the disturbance (matter particle), present in the gap. Such an effort ceases only when the disturbance is removed from the 2D energy field and the continuity of the 2D energy field is restored. A 2D disturbance/body co-exists with 2D energy fields in all planes other than in its own plane. A 2D disturbance does not create discontinuity in 2D energy fields of other planes. Consequently, a 2D disturbance is acted upon by the gravitational force/pressure produced only by the 2D energy field in its own plane. Since a quantum of matter, in its 3D state, occupy space in more than one 2D energy field, all the 2D energy fields in the planes passing through the 3D quantum of matter act on it.

Attempt by the quanta of matter of a 2D energy field, to extend or move axially, applies the gravitational effort. If such axial movements are not permitted by any means, gravitational effort cannot act on the disturbance but the effort remains as long as the disturbance is present. We may say that after the stabilization of a disturbance, the gravitational effort become inactive but remains being applied. [A force comes into being only when there is a motion. As such, term 'force' is often substituted by term 'effort' in dynamics]. Magnitude of the gravitational effort depends on the extent of the 2D energy field in the direction from where the effort is being applied. Its dependence on the angle subtended by the 2D energy field is explained in another section of this book and the effect of curvature of perimeter of the disturbance is considered later in this section. Since the gravitational effort is effective from any distance, up to the limit of extent of the 2D energy field, which extends to infinity in free space, gravitational action is a long-range force. That is, the gravitational action in any direction, felt at a point, is applied by the combined action of all the quanta of matter in a 2D energy field latticework, in that direction. It is the total number of quanta of matter (extent of 2D energy field) acting on the disturbance, which determines the magnitude of gravitational action (within certain limitations) on a point on the periphery of a disturbance.

Unlike the gravitational action, all other field forces are

produced due to angular displacement (with or without an axial movement) of the constituent quanta of matter in a 2D energy field. There need not be a discontinuity of the 2D energy field. Field forces are reactions applied by quanta of matter of 2D energy fields, trying to regain their stable position within the quanta-chains of a 2D energy field. Different manifestations of field forces may be associated with each other. They may appear and act on or about a body simultaneously. During apparent interactions involving 3D matter particles, gravitational action is always present. Because, 3D matter particles are disturbances with respect to 2D energy fields and it is the gravitational pressure, which creates and sustains 3D matter particles. All 2D disturbances and 3D matter particles (and matter-bodies in higher dimensional systems, if any) break the continuity of 2D energy fields of their existence. They exist within the gaps formed in the 2D energy fields.

3.3. Action of gravitation:

Action of an effort produces work about a body. We shall consider only that part of work-done, which changes the state of motion of a body. Rate of change of state of motion is the force. Work can be done only by having motion and displacement (a change of state of motion) of the body. (Here, only the movements of quanta of matter within the 2D energy fields are considered). Variation of matter content of a particle does not constitute work-done (hence, heating or cooling a body is not work-done; they vary the matter content and size of a body). Motion or displacement of a body may be that of the whole body or of a small part of the body. Therefore, a force essentially requires a movement of the point of application of an effort. State of motion of the body has to undergo a change. Otherwise, the applied effort can be considered as inactive. When it is inactive, applied effort is continuously neutralized by the reactive effort (or any effort in the opposite direction) and hence it cannot produce resultant work about a body.

To do work, the point of application of an effort has to move in relation to other parts of the same body or system. If no movement can take place, no work is done and it may be considered that, though the effort is being applied, it is not acting. No force is generated. This is the same as stating that no energy is transferred and hence no work is done. Work-done on a matter-body is the changes in the 2D energy fields latticework squares associated with the body, produced by the movement and displacements of quanta of matter in it. As long as an effort cannot introduce movements of the quanta of matter in the 2D energy field-latticework squares, associated with the body, it cannot be considered to be acting on the body. Since the 'energy' is related to the work-done, action of an effort means a transfer of work-done from the region of one body to the region of another or from the region of one part of the body to the region of another part. Mere presence of an effort (application of force) does not constitute action of a force.

During the formation of a 2D disturbance, its matter density (or quantity of quanta of matter per unit area) being low, the gravitational effort generated in the 2D energy field produces axial movements of quanta of matter of the latticework towards the (center of the) disturbance. Such movements of the quanta of matter, into the space of the gap, shrink the gap in the 2D energy field and reduce the size of the disturbance, contained within the gap. During the size-reduction of the disturbance, space/area occupied by the disturbance is reduced along with its perimeter, in its plane. Reduction in its size reduces the disturbance's perimeter, which is its magnitude. Reduction in size of a disturbance increases the magnitude of the gravitational force on it, due to the increase in the curvature of its perimeter. Reduction in the magnitude of a disturbance does not change its matter content, but increases its matter density. Unless additional quanta of matter are added to the disturbance, its matter content remains constant. It is the variations, in the extent of empty space between adjacent quanta of matter in the disturbance, which determines the matter density of a disturbance. ['Empty space', mentioned here indicates absence of quanta of matter in certain parts of the plane of disturbance only. This 'empty space' remains filled with constituent quanta of matter of 2D energy fields in other planes.] Larger gaps between quanta of matter of a disturbance produce lower matter density. When there are no empty spaces between the quanta of matter in a disturbance, in its plane, matter density of the disturbance will be equal to that of a quantum of matter.

In a disturbance, the constituent quanta of matter are not arranged in any order but they are gathered together at random. Each of these constituent quanta of matter is constantly trying to grow in its own spatial dimension and thereby tries to occupy more space in its spatial dimension. Haphazard expansion of quanta of matter creates gaps in between them in the plane of the disturbance. Action of the gravitational effort is against this tendency of the quanta of matter to expand the space of the gap in the 2D energy field. Gravitational pressure confines the quanta of matter of the disturbance to the reducing space of the gap. Gradual reduction in its size increases disturbance's matter density. This process will continue until the matter density of the disturbance has reached a 'maximum limit' in 2D space system. At maximum matter density of the 2D disturbance, all quanta in it are transformed into their 2D spatial state in the plane and they are brought to stay in contact with each other on all sides. There are no gaps in the plane between the quanta of matter in the disturbance. Quanta in the middle regions of the disturbance alter into appropriate geometrical shape to fit with each other without gaps between them. Gaps between quanta of matter reduce matter density of the disturbance.

Rise in the matter density of a disturbance is made possible by the action of gravitational effort from all around the disturbance. Energy expended (measure of a functional entity, which represents the stress of quanta of matter in the 2D energy field around the disturbance due to total resultant motion of quanta of matter around the disturbance) by the 2D energy field is stored in the disturbance in the form of pressure energy. The same is also stored in the 2D energy field as its stress due to distortions. At maximum matter density (in 2D space system), matter content within the disturbance can no more be shrunk but the effort by the 2D energy field continues to be applied on the disturbance from opposite directions (on either side of disturbance). Further movements of quanta of matter of the 2D energy field towards the (centre of) disturbance are possible only if part of matter content in the disturbance is removed from the plane.

In a hypothetical case, where removal of matter content is not permitted from the plane; at this stage, gravitational actions on the disturbance come to a halt. That is, though the gravitational effort is present, it cannot produce further movements of quanta of matter of the 2D energy field towards (the centre of) the disturbance and hence it is not acting on the disturbance any more. No more work is done on or about the disturbance. Hence, energy transfer does not take place from the 2D energy field into the disturbance. Though the gravitation is not active anymore, it is always present on the disturbance and the effort of the gravitation, to act on the disturbance, is continuously maintained. Gravitation can act again on the disturbance only when, for any reason, axial movements of quanta of matter of quanta-chains in the 2D energy field are permitted in their spatial dimensions, towards the disturbance. Reasons for such permission may be a reduction in the matter density of the disturbance by loss of matter content or a reduction in the magnitude of gravitational effort applied on the opposite side of the disturbance.

In the case of a reduction in the matter density (internal pressure) of a disturbance, quanta of matter in the quanta-chains of the 2D energy field from all around the disturbance are permitted to move towards (the centre of) the disturbance, whatever is the condition of gravitational effort on the opposite side of the disturbance. Gravitational effort is now acting on the disturbance and if equal magnitude of gravitational effort is also acting from opposite direction, they will jointly reduce the space occupied by the disturbance and thereby compress the disturbance back to its maximum matter density in the 2D space system. Gravitational pressure on the disturbance stops its action when the matter density of the disturbance, once again, reaches its maximum value in the 2D space system. After this, the gravitational effort continues to be applied on the disturbance but it remains inactive. This action maintains the matter density of a stable disturbance, automatically at a constant level.

Magnitude of gravitational pressure, on a disturbance, depends also on the curvature of its perimeter. If the curvature is relatively small, magnitude of gravitational pressure, on a disturbance is much lower. It can be neutralized by a small increase in the matter density (internal pressure) of a disturbance. Hypothetical maximum matter density mentioned above, without taking the curvature of disturbance's perimeter into consideration, is only a theoretical concept to explain the actions of gravitational pressure on a disturbance. A balance between internal pressure of a disturbance of smaller size and the gravitational pressure, acting externally on it, is achieved by a reduction in the gravitational pressure around the disturbance due to 'jamming effect' of 2D energy fields.

During the motion of quanta of matter in the quanta-chains, towards the disturbance, latticework squares in the 2D energy field are distorted. Latticework squares nearer to the disturbance have greater distortions. They are also compressed to a greater extent compared to latticework squares farther away from the disturbance. If matter density of the disturbance is very low, 2D energy field will attempt to reduce the area of the gap until the matter density reaches its highest magnitude. For this to happen, latticework squares of 2D energy field have to be compressed or distorted to very large extent. This may not always be possible due to proximity of neighboring latticework squares. The latticework squares of 2D energy fields cannot be compressed by gravitational actions beyond certain limit. As and when this limit is approached, magnitude of gravitational actions reduces. Gravitational action on the disturbance stops altogether, when the limit is reached. This is the 'jamming effect' of 2D energy field.

If the matter density of a disturbance is very low, 2D energy field will shape the disturbance into a circular object. Latticework squares of the 2D energy field will settle around the disturbance in a circle, whose diameter depends on the nature of 2D energy field, which is usually steady and suitable for present state of the universe.

Action of gravitational pressure on a disturbance compresses the disturbance. While doing so, the latticework squares in the 2D energy field nearer to the disturbance also get compressed. Latticework squares nearer to the disturbance achieve higher degree of distortions than the latticework squares farther from the disturbance. Development of distortions, around the disturbance, produces an area (a volumetric space in case of 3D disturbances/ particles) of distortions (distortion field) about the disturbance. This region around the disturbance is the 'gravitational field' about the disturbance. Gravitational field is the distorted region of 2D energy fields around a 3D matter (macro) body. Depending on the nature of movements of basic 3D matter particles, gravitational fields around them may be transformed into different types of (electromagnetic and nuclear) fields, used in the phenomena of natural forces. All natural and derived forces originate from gravitational efforts of 2D energy fields.

3.3.1. Motion of a particle by gravitation:

Should the gravitational effort, on any one side of a disturbance diminish, matter density of the disturbance is bound to come down. This facilitates axial motion of the quanta of matter of quanta-chains in the 2D energy field latticework from the opposite direction, towards the disturbance. Gravitational efforts become active on the disturbance at once. Action of the gravitational effort attempts to increase the matter density of the disturbance by moving the side of the disturbance inward. Since the gravitational effort applied on the opposite side of the disturbance is maintained inactive (or magnitude of its action on the opposite side is reduced), this attempt does not result in raising disturbance's matter density.

As long as magnitude of the gravitational force on the opposite side of the disturbance is kept lesser, this action of gravitational effort to push at the disturbance continues. Gravitational action on one side of the disturbance continues to move that side of the disturbance in the direction of the effort. Gravitational action on the disturbance, in an effort to compress it (if the disturbance is large enough), in addition, produces a linear motion of the disturbance in the direction of the effort, by parting the latticework fabric of the 2D energy fields and passing the disturbance through the gap created. If the effort is not large enough, the disturbance will move only so much as is required to create enough reaction from the 2D energy field to oppose the gravitational effort and stop its action on the disturbance, to move it. Stopping the disturbance will help to raise its matter density and internal pressure. This is how a gravitational effort produces the motion of the disturbances (matter particles) in space.

When there are two disturbances, present in a plane, gravitational effort on them from in between the disturbances is always less than the gravitational efforts on them from outer sides. Differences in magnitudes of gravitational efforts move the disturbances towards each other. This appears to us as the disturbances are under some kind of mysterious mutual attraction. This phenomenon gives rise to the 'apparent gravitational attraction' between the disturbances. In reality, each of the disturbances is moved independently by separate gravitational actions by the 2D energy field on them. Motions of matter-bodies under apparent gravitational attraction are also inertial actions. In nature, apparent gravitational attraction is relatively a minor by-product of gravitational actions. Gravitational efforts mainly act to create and sustain disturbances and matter-bodies in higher dimensional space systems.

If the matter density of a disturbance can be kept constant at the maximum limit, by some means, irrespective of the condition of the force on its opposite side, a gravitational effort cannot act on a disturbance. This is because no axial movements of the quanta of matter of quanta-chains in the 2D energy field are permitted towards (the centre of) the disturbance. Even if the gravitational effort on the opposite side of the disturbance is totally absent, gravitational effort cannot act on a disturbance, whose matter density is at maximum level. (This is a hypothetical case, disregarding the balance between internal and external pressures). There can be no inertial motion due to gravitation, of the disturbance, whose matter density is maintained at the highest level. Thus, it is seen that the action of (work done by) the gravitational effort takes place only when certain axial movements of the quanta of matter of quanta-chains in the 2D energy fields are permitted towards the disturbance, in the direction of the effort.

Contrary to present beliefs, gravitational actions are neither universal nor constant. Its actions may be limited or vary under definite laws. Actions of gravitational effort, including apparent gravitational attraction, depends on many factors like; surface shape of basic 3D matter particles, gravitational efforts applied from different directions, nature of 2D energy fields in a region, etc. Inertial motions of disturbances, whose matter density is maintained at maximum level, are produced due to their shape rather than a change in their matter density, as explained below.

3.3.2. Pressure energy of a disturbance:

Conventionally, during a work being done, energy (measure of a functional entity denoting the ability to do work, which represents the total resultant stress produced by the displacement of quanta of matter in the 2D energy field about the disturbance) is expended. By the concept explained in 'Hypothesis on MATTER', work is the primary real entity, which creates other functional entities like, force, power, energy, etc. To do work in a macro body, by inertial action, work-done in another macro body is transferred into it. Force-applying body expends work and the force-receiving body receives the same work. In case of basic 3D matter particles, work is done by the 2D energy fields within themselves to create and sustain these particles. Energy is the resulting stress in the 2D energy fields and in the quanta of matter associated with the matter-bodies/particles. During the compression stage of a disturbance, gravitational action does the work. Related energy is stored in the form of displacements and reduction in the length of the quanta of matter, both within and outside the disturbance. These quanta of matter are held at shortened length against their natural tendency, to expand in their own (single) spatial dimension. This, in turn produces stress within the quanta of matter and in the latticework structure formed by them. Stress in the shortened quanta of matter of the disturbance is the energy stored within the disturbance as 'pressure energy'. Certain energy is stored in the 2D energy field around the disturbance, as the stress produced due to its distortions, required to sustain the disturbance.

When a 2D disturbance has reached its highest matter density and the compression of the disturbance in 2D space system is completed, gravitation cannot act on it any more (unless, there is an imbalance between external and internal pressures or loss of matter content from the plane of the disturbance). Hence, the work, which was being done to compress the disturbance, is stopped. No more additional work can be transferred into or about the disturbance and the magnitude of energy stored in and about the disturbance in the form of pressure energy or stress due to displacements of quanta of matter remain constant.

Energy, already stored in and about a disturbance, remains

with the disturbance as long as its matter density (internal pressure) remains unchanged. When the gravitational action is able to produce an inertial movement of the disturbance, work-done during the acceleration of the disturbance is stored about the disturbance in the form of additional distortions in the 2D energy fields in its immediate neighborhood. Stress produced by the moving distortions in the 2D energy field is the kinetic energy associated with the moving disturbance. During the motion of the disturbance, part of the gravitational effort, producing the inertial movement, acts as an inertial force. In all these cases, the gravitational action, which is derived from the inherent property of 2D energy field, to maintain its continuity and stability, provides the basis for all types of forces and energy.

3.3.3. Gravitation on a straight perimeter:

Let us consider a 2D disturbance at its highest matter density and has part of its perimeter as a straight line. In case of a 3D disturbance, this is analogous to partially flat surface. A disturbance reaches its highest matter density, when all its constituent quanta of matter are fully converted into the present (2D or 3D) dimensional space system. In this state, neighboring quanta of matter in the disturbance fit close with each other and there are no gaps or empty spaces between the constituent quanta of matter in the plane of the disturbance. Highest matter density is the same as the matter density of a quantum of matter in its free state. Variations (reductions) in the matter density of a disturbance occur due to presence of gaps in between its constituent quanta of matter. Since the disturbance has reached highest matter density, gravitation from the 2D energy field is no more active on it (though the gravitational effort is continuously applied). 2D energy field and the disturbance are in stable equilibrium state.

Consider part of a 2D energy field as shown in figure 3C. Arrows in bold lines (perpendicular to each other) show the gravitational action through the quanta of matter of quanta-chains in 2D energy field. Part below the dashed lines shows part of a disturbance. Arrows in grey dashed lines show the resolved components of gravitational efforts, applied through the quanta of matter. Presence of a disturbance in a 2D energy field causes a discontinuity in it. Because of the discontinuity, 2D energy field applies gravitational effort onto the disturbance. Arrowheads on the quanta of matter show the directions of the gravitational effort, applied through them. Because of the straightness of the perimeter (shown by the double line) of the disturbance, junction points of the 2D energy field, in contact with the disturbance, at this part of the perimeter, are in straight line and all related latticework squares of the 2D energy field are in neutral and stable condition. For a gravitational effort, to act on to a disturbance, axial movements of quanta of matter in the 2D energy field latticework are required.

From the figure 3C, it can be seen that the gravitational efforts, applied by the 2D energy field, are along the chains of the quanta of matter in them. Directions of all these efforts are at 45° to the straight perimeter of the disturbance. Each of these efforts may be considered to have two resolved components, shown by arrows in grey dashed lines. One component, perpendicular to the perimeter of the disturbance adds up with similar components from other efforts and applies directly into the disturbance (perpendicular to the perimeter). As and when the matter density of the disturbance reaches highest limit; opposite effort from the internal pressure of the disturbance balances this component of the gravitational effort. Vertical components of gravitational efforts (applied through all quanta of matter) are equal and unidirectional. Hence, the compressive force applied on the straight periphery of the disturbance is uniform and even. Uniform internal pressure of the disturbance balances these components of gravitational efforts.

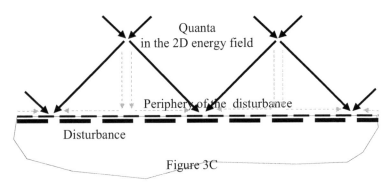

Figure 3C

Other components of all gravitational efforts are parallel to the straight section of perimeter of the disturbance. These components, being in opposite directions, neutralize each other. This prevents relative motion between quanta of matter of the 2D energy field. Consequently, as long as the internal pressure balances the vertical components of gravitational action, applied into the disturbance, there will be no resultant gravitational action on the disturbance, at this perimeter section.

If the matter density of the disturbance is now lowered, due to lower internal pressure, balance between vertical components of gravitational efforts and vertical components of reactions due to internal pressure at the straight perimeter of the disturbance is offset. Quanta of matter in the 2D energy field are permitted to move towards the disturbance. Their displacements towards the disturbance are able to increase the compressive force on the disturbance. Thus, the gravitational effort is able to act on the disturbance. Motion of the quanta of matter towards the disturbance tends to increase disturbance's internal pressure. However, if the matter density of this disturbance can be maintained, by some means, at the highest constant level at the straight perimeter; under any condition, the quanta of matter of the 2D energy field cannot move towards the disturbance to cause gravitational action on it. At the perimeter of the disturbance, the gravitational action by the 2D energy field is balanced by equal and opposite reaction applied by the internal pressure of disturbance. This is true even if the gravitational action on the opposite side of disturbance is totally absent.

Above-mentioned occasions are the only instances when a disturbance (or 3D matter particle) does not invoke gravitational action. There is no action by gravitation on a straight perimeter (a flat surface in case of 3D particle) of a 2D matter-body. In nature, all basic 3D matter particles are disc shaped, with almost flat faces. These disturbances have (almost) flat perimeter at their faces and curved perimeter at their circumferences. Their matter density is maintained at highest level. Therefore, gravitation can act on them only in the disc-planes containing their circumference. Gravitation is unable to act on these disturbances in any other planes passing through their faces, as long as the matter density is maintained at the critical level. Gravitation is able to act on their disc-faces, when changes in the internal pressure vary the curvature of their flat faces. This action helps to maintain the basic 3D matter particle's shape and matter density.

This phenomenon is not applicable to composite 3D macro bodies. All macro bodies are constituted by numerous basic 3D matter particles. Gravitational actions take place only on the basic 3D matter particles, because they break the continuity of 2D energy fields. Each 3D matter particle in a macro body is affected by gravitational actions separately. Resultant of gravitational actions on all 3D matter particles of a macro body, together, constitutes the apparent gravitational action on a macro body. Therefore, the shape of perimeter or matter density of a macro body does not affect gravitational actions on it.

In a stable and undistorted 2D energy field, all junction points are in straight lines. Hence, the above explanation is suitable only for those perimeter sections of a disturbance that are parallel to the line joining the junction points in the 2D energy field. Should the perimeter section of the disturbance be at an oblique angle to the line joining the junction points of a stable 2D energy field, 2D energy field latticework squares distort themselves so that their junction points are in contact with the perimeter-section. Wherever such contact cannot be established, the junction point will stay away from the straight line and the disturbance's perimeter will extend towards the junction point to make contact. Thus, the perimeter section will no more be a straight line with respect to the 2D energy field. When we say, the 2D energy field is applying a force on a disturbance, contacts between the quanta of matter of the 2D energy field and the disturbance are at the junction points of quanta of matter. A 2D energy field can act only through its junction points. Junction points of a 2D energy field come to rest on the perimeter/surface of a disturbance and apply forces directly on to the (condensed) 3D matter-body, made up of quanta of matter within the disturbance. Resultant gravitational action on a straight perimeter by a 2D energy field, line joining whose junction points are oblique to the perimeter, tends to turn the disturbance perimeter so that it becomes parallel to the line joining the junction points of 2D energy field.

3.3.4. Gravitation on a curved perimeter:

Consider a perimeter-section of a disturbance as shown in figure 3D. It has a convex curvature. Arrows in bold lines, forming the rectangles, show the gravitational actions through the quanta of matter of quanta-chains in the 2D energy field. Part, below the dashed curved line, shows part of a disturbance with curved perimeter. As it can be seen from the figure, in stable equilibrium condition, 2D energy field-latticework squares are not at their neutral state. They are strained. Quanta of matter of the 2D energy field latticework square in the middle are compressed and the square is deformed towards the disturbance. Quanta of matter of the latticework squares on either side of centre point are in tensile state. Arrows in bold lines represent directions of gravitational actions through each quantum of matter. Arrows in grey dotted lines show resolved components (they are not to scale) of these actions. Lateral components of all gravitational actions, being equal and opposite, neutralize. There is no resultant lateral effort applied on the disturbance.

Vertical components of gravitational actions are not equal. Depending on the state (tensile or compressive) of quanta of matter in the 2D energy field, resultant actions are either towards the disturbance or away from it. Gravitational actions of different magnitudes, applied through various quanta of matter, produce different resultant vertical efforts at various junction points. Internal

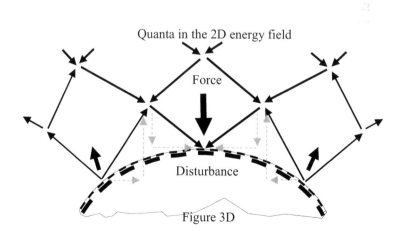

Figure 3D

pressure of the disturbance may be taken as uniform. Differences in the applied efforts produce different resultants at these junctions. Resultant effort from latticework squares presses at the middle point of the disturbance's perimeter as shown by the thick arrow. Resultant efforts at the junctions, away from the middle point, are such as to reduce the compressive pressure on the disturbance. In the case, as shown in the figure, middle point of disturbance's perimeter is pushed downwards and its outer sides of the disturbance are pulled (comparatively) upwards. Tendency of resultant gravitational effort on the curved perimeter is to straighten the perimeter. Changing the shape of the perimeter to a straight line will restore the neutrality and stability of the 2D energy field.

Hence, there is an additional gravitational effect acting on the curved perimeter-section of the disturbance towards the middle of its convex curvature. This additional gravitational effect, applied on the disturbance, is over and above the normal gravitational action, which is balanced by the internal pressure of the disturbance. Geometrical shape of the disturbance's perimeter is the only cause for the additional gravitational effect. Since the matter content of the disturbance is already at its highest matter density, this additional gravitational effect can act on the disturbance to produce linear motion of the whole disturbance or its part.

As long as equal and opposite gravitational effort is applied on the opposite side of the disturbance, this effort is neutralized and the disturbance maintains a steady state of motion, without linear displacement of quanta of matter in contact. Irrespective of the condition of internal pressure (matter density) of the disturbance, action of this part of gravitational effort depends on similar effort being applied on the opposite side of the disturbance in the same plane. Magnitude of inertial motion, imparted to a disturbance with convex curved perimeter, solely depends on the balance between the gravitational efforts, applied on either side of the disturbance. Inertial motion imparted, by gravitational action, to a stable disturbance with different curvatures on opposite sides is in a direction from higher convex-curved side to lesser convex-curved side. Therefore, a stable disturbance with different curvatures on opposite sides remains always under inertial motion, irrespective of its state of compression.

In nature, it is rare for a stable disturbance to have a concave part of its perimeter. Gravitational actions on concave perimeter-section is similar but in opposite direction to that described above. It is the tendency of the gravitational action to straighten out any curved perimeter-section of a disturbance. There are many 2D energy fields in contact with a flat surface of a 3D disturbance. Not all of them are perpendicular to the surface. In those 2D energy fields, which are at oblique angle to the surface, there are only few junction points, which are in contact with the disturbance. All others stay at a distance from the surface of disturbance and are under the action of reactive forces from their neighboring quanta of matter. Therefore, as far as the gravitational action by any of such 2D energy field is concerned, its magnitude is very little and can be neglected.

It can be concluded that: provided the matter density of a disturbance is kept constant at the highest level, in 2D/3D space systems, the gravitation cannot act on straight perimeter (flat surface in 3D disturbance) of a disturbance. Magnitude of gravitational action on the disturbance by a 2D energy field depends (also) on the magnitude of convex curvature of the perimeter of a disturbance. Let this relationship be represented by a term K_2. Therefore, the equation (3/1) for the gravitational force, GF, acting on a point on the perimeter of a disturbance has to be modified accordingly and becomes:

$$GF = \frac{G_1 K_2}{2\pi d} \qquad (3/2)$$

In nature, we usually come across only 3D macro bodies, which are made up of numerous basic 3D matter particles. Circumferential curvatures (radial size) of all basic 3D matter particles, in their stable states, are the same. Hence, the constant of proportion, K_2, in the equation (3/2) is of little interest. We need not consider it separately. It merges with the gravitational constant G_1.

In the above discussion, it was seen that for gravitational effect to act on a disturbance, it is necessary for the quanta of matter in the 2D energy field to move in relation to each other and towards the disturbance. This is true in cases of all field forces.

Should the disturbance itself is moving at certain speed in any linear direction; effect of the effort, in the direction of motion of the disturbance, is correspondingly reduced. In order to have the same work done on a moving disturbance, quanta of matter, applying the (gravitational) effort need to move faster. In other words, to have the same effect (in terms of acceleration) on a moving disturbance, as that on a static disturbance, a greater (gravitational) effort has to be acting on the disturbance. This phenomenon, gives rise to the phenomenon of 'relativistic mass' (increase in mass in proportion to the linear speed of the disturbance). It can be stated here that the magnitude of (gravitational) effort, on a moving disturbance (body/particle) in the direction of its motion, depends also on the linear speed of the disturbance (body/particle) in the direction of gravitational action. If the linear speed of the disturbance approaches the speed of light, an external effort will be unable to act on a disturbance. Inability of the external effort to act on the disturbance, moving at the speed of light, is presently assigned to an assumed conversion of energy into mass with a corresponding increase in the mass of the disturbance to infinity.

3.4. Apparent gravitational attraction:

By its inherent property, a 2D energy field tends to reduce the size of a disturbance in it, to minimum magnitude. One way to do this is by compressing the disturbances to their minimum radial sizes and to make them circular in shape. As the size of a disturbance is reduced and its shape made circular, its perimeter, which is the magnitude of disturbance, is also reduced. For any given area, the circular perimeter is the least. It was seen in section 3.2.1 that immediately on forming a disturbance; its shape is made circular by the action of gravitational pressure. Now, the perimeter of a larger circle is lesser than the sum total of perimeters of smaller circles whose combined area is equal to the area of the larger circle. Consequently, another way to reduce the total magnitude of disturbance in a 2D energy field is, for it to combine the disturbances in it, to make a larger but single disturbance. This is achieved as follows:

When a disturbance is by itself, in a 2D energy field, gravitational action on it from all around is such as to confine it in shape and size. Magnitudes of the gravitational actions, onto a disturbance in free space, from all directions in the plane are equal. This is because the extent of 2D energy field in any direction, from a single disturbance in free space, is infinite. When there are more than one disturbance in the same 2D energy field plane, extent of the 2D energy field in between them is equal to the distance between them. Gravitational actions by this smaller extent of 2D energy field, on these disturbances are lesser than the gravitational actions on them by infinite extent of 2D energy field on their outer sides (sides away from each other). Imbalance in the magnitudes of the gravitational actions on these disturbances – higher on the outer sides and lower on the inner sides – compels them to move towards each other.

In practice, matter particles are observable and the 2D energy fields remain hidden to us, the 3D beings. Hence, this action appears to us as if it is taking place between the disturbances (matter particles). While, they are actually pushed towards each other, they appear to be pulling each other. Hence, this action is called 'attraction due to gravitation' between these disturbances. This phenomenon causes the inertial motion of both disturbances towards each other. Since, no logical cause for such an action is available, it is assumed to occur under 'apparent attraction' attributed to masses of the disturbances, due to gravitation.

Magnitude of the apparent attraction due to gravitation between two disturbances depends on the gravitational actions on each of the disturbances (which depend on their sizes, curvature of perimeter and angle subtended between them) and the distance between them. Distance between the disturbances comes into play due to the extent of the 2D energy field separating the disturbances. Magnitude of gravitational action on larger disturbance is greater than the gravitational action on a smaller disturbance. Larger disturbance has larger contact perimeter and smaller disturbance has smaller contact perimeter with a 2D energy field. While considering the apparent gravitational attraction, higher push is applied on larger disturbance towards the smaller disturbance and smaller push is applied on smaller disturbance towards the larger disturbance. In case of macro bodies, this difference in magnitude is not noticed for different reasons, as explained later in this book. Since a 2D energy field applies the gravitational effort, directly onto the disturbances, (apparent) attraction due to gravitation or gravitational apparent attraction does not require any other medium of transmission or a carrier. In fact, apparent gravitational attraction is an apparent manifestation of the effect of gravitation on each disturbance, separately, by the 2D energy fields.

Hence, in the case of apparent gravitational attraction, there is no force transmitted from one disturbance to another, but each disturbance is acted upon by gravitation separately to move them towards each other and thereby to produce a combined apparent effect of the disturbances being pulled towards each other. It is the difference in the magnitudes and directions of the gravitational actions on them, which brings them towards each other. There are no direct interactions between the disturbances. No imaginary particles or no actions at a distance are required to produce apparent gravitational attraction between two disturbances. Gravitational efforts act (or are applied) on the disturbances continuously and separately to cause the apparent attraction between them. Hence, the apparent gravitational attraction between two disturbances/bodies is modified instantly on changes of parameters of any one or both the bodies. This instantaneous modification of apparent attraction due to gravitation takes place at any distance between them, however large it may be. Hence, supposition of a gravitational field around a body (unless it means the distorted region 2D energy field around a macro body) or virtual particles like gravitons (that may travel faster than light) are not necessary to explain the phenomenon of apparent gravitational attraction. Magnitude of apparent gravitational attraction is subject to instantaneous modification, between matter-bodies in nature.

Apparent gravitational attraction between two disturbances tends to move them towards each other. If the apparent attraction is small, each disturbance will move towards each other until reaction from the 2D energy field-latticework structure (on the inner side) balances the apparent attraction. Thereafter, the disturbances will maintain their distance. In case the apparent gravitational attraction is strong enough to move the disturbances by parting the latticework of the 2D energy field, the disturbances will move towards each other until they merge. As and when such movements take place, work is done to displace quanta of matter in the 2D energy field-latticework and energy is developed in the strained 2D energy fields. Until the displacements of quanta of matter in the 2D energy field are permitted, such work could not be accomplished. During the motion of a disturbance, introduced by the apparent gravitational attraction, substance of the disturbance (matter content of a body) does not undergo any change. Work, created by the 2D energy field in producing the motion, remains within the 2D energy field about the disturbance, and continues to produce disturbance's motion at constant rate. We may say that the work expended by the gravitation is now stored in association with the disturbances (in the 2D energy field), to produce kinetic energy of the disturbances. As the work-done (distorted region of 2D energy field around the disturbance) transfers itself in the 2D energy field, it carries bodies of the disturbances, along with it.

Work remains in association with the disturbances in an intangible form (because, we cannot measure movements or displacements of quanta of matter in the 2D energy fields) as distortions to 2D energy fields' latticework squares. Apparent attraction due to gravitation between two point-bodies in a 2D energy field is limited within the plane containing both the point-bodies. If the disturbances are in different planes, there is no apparent gravitational attraction between them.

3.4.1. Magnitude of apparent gravitational attraction:

In figure 3E, let 'A', 'B' and 'C' be three points in a straight line in a 2D energy field plane. Let 'A' and 'B' be two circular 2D point-disturbances of unit measure each.

Let the distance: $AB = BC = d$.

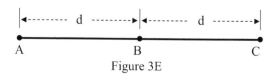

Figure 3E

Consider the gravitational effort on the point-disturbance 'B' along the line AC. Gravitational effort on disturbance 'B' along CB by the 2D energy field of infinite extent is equal to the maximum value of constant G_1. If there are no other disturbances in the same plane, an effort of equal magnitude would be acting on disturbance 'B' along AB as well. Gravitational effort on a point-disturbance is applied from all around the point. However, due to the presence of a disturbance at point 'A', the extent of 2D energy field along BA is limited up to the point 'A'. Consequently, the gravitational effort on point-disturbance 'B' along AB is produced by the part of 2D energy field whose extent is from point 'B' to point 'A' only.

Gravitational effort on disturbance B by the extent of 2D energy field from point 'B' to point 'C', is equal but in opposition to the gravitational effort on point 'B' by the extent of 2D energy field from point 'B' to point 'A'. Therefore, it can be taken that they neutralize each other. They can only compress the disturbance at point 'B'. (While considering the apparent gravitational attraction, we are interested only in linear inertial actions of the disturbances, so as to move them towards each other. Hence, the compressive actions of gravitational efforts are ignored). Resultant gravitational effort on point 'B' along CB, is the gravitational effort on it by the 2D energy field beyond point 'C', onto the right. Since we are considering (circular) point-disturbances, we may neglect other factors like angle subtended, curvature of perimeter, etc.

Magnitude of the gravitational effort on point 'B' through point 'C',

$$F = \frac{G_1}{2\pi} \times \frac{1}{d} \qquad \text{by equation (3/1)}$$

A similar gravitational effort is applied on the point-disturbance 'A' also. Hence, the total magnitude of gravitational effort on these disturbances to move them towards each other is the (apparent) attraction due to gravitation.

Magnitude of apparent gravitational attraction between disturbances 'A' and 'B',

$$F = 2 \times \frac{G_1}{2\pi} \times \frac{1}{d} = \frac{G_1}{\pi d} \qquad (3/3)$$

Let these disturbances be of larger sizes. Apparent gravitational interaction between large 2D disturbances takes place between every point (of unit measure) on their perimeters, within the angle subtended by these disturbances on each other. Number of such points on each disturbance is proportional to the magnitude of the disturbance, larger disturbance having more number of points and smaller disturbance having lesser number of points on their circumferences. For the time being, we may neglect this relation and take magnitudes of the disturbances to represent the number of these points. Length of its perimeter is the magnitude of a disturbance in a plane. Let magnitudes of these disturbances be 'm_1' and 'm_2', respectively. Each point on the perimeter of one disturbance interacts with each of the points on the perimeter of second disturbance. Although, it is the 2D energy field that is acting on each of the disturbance separately, for simplicity of the explanation we may consider that each part of a disturbance interacts with every part of the other disturbance, to produce apparent gravitational attraction.

Magnitude of the first disturbance = m_1

Magnitude of the second disturbance = m_2

Total magnitude of disturbances, interacting to produce apparent attraction = $m_1 + m_2$

Every point in one disturbance interacts with every point in the other disturbance. Total number of interactions is given by the multiplicand of the magnitudes of disturbances.

Number of interactions = $m_1 \times m_2$

Multiplying equation (3/3) by number of interaction, we get the total magnitude of apparent attraction, F, due to gravitation between the disturbances,

$$F = \frac{G_1}{\pi d}(m_1 \times m_2) \qquad (3/4)$$

Part of this apparent attraction, F, contributed by each of the disturbance, is proportional to its magnitude.

Magnitude of gravitational effort (contributing to apparent attraction) on the first disturbance,

$$F_1 = \frac{G_1}{\pi d}(m_1 \times m_2) \times \frac{m_1}{(m_1 + m_2)} \qquad (3/5)$$

Magnitude of gravitational effort (contributing to apparent attraction) on the second disturbance,

$$F_2 = \frac{G_1}{\pi d}(m_1 \times m_2) \times \frac{m_2}{(m_1 + m_2)} \qquad (3/6)$$

Greater part of this apparent attraction due to gravitation is the share of the larger disturbance. Magnitude of a disturbance in a plane, being equal to its perimeter, larger disturbance has greater magnitude. This type of sharing of the apparent attraction by gravitational efforts can be considered only when we are considering one or few basic 3D particles in space. Sharing of apparent attraction in proportion to their matter content is not applicable to 3D macro bodies. (See section 5.1.4). Even a very small 3D macro body contains millions of basic 3D particles in different phase relations. At any instant, only very few of these particles contribute towards the apparent gravitational attraction between the macro bodies.

When we consider total apparent gravitational attraction between two 3D macro bodies, to be a single 'attractive force' between them, apparent gravitational attraction on each macro body due to the other is never considered separately. When the disturbances are larger than point-bodies, other factors affecting the gravitational effort, on a disturbance, also should be considered. Putting the constants of proportion explained in the last section is of theoretical interest only. In 3D world, all disturbances (basic 3D matter particles) are almost of uniform size and shape. All other macro-bodies are union of the basic 3D matter particles in various combinations. Gravitational actions are onto the constituent basic 3D particles of a macro body rather than to the composite body.

Apparent inertial action due to gravitation, between two bodies, is termed 'attraction', because the resultant of the gravitational actions, acting separately on each of the disturbances, tends to move the disturbances towards each other. Gravitational action has a push effect on the disturbances rather than a pull effect. There are no pull effects emanating from the disturbances. Gravitational actions push each of the disturbances towards each other. There are no direct actions between the disturbances. Two disturbances moving towards each other under separate push-forces on them appear to be attracting each other. The fact that the extent of 2D energy field, in between two disturbances, is always less the extent of 2D energy fields on their outer sides forbids 'apparent repulsion' between two disturbances due to gravitational actions. Extent of 2D energy field is considered only with respect to the disturbances considered. Presence of other disturbances in the same plane or same line does not make any difference, as explained later in this book.

3.4.2. Effect of angle subtended:

From the equation (3/4), it is seen that the magnitude of apparent gravitational attraction between two disturbances in a 2D energy field plane is proportional to the product of their magnitudes and inversely proportional to the distance between them. All other variable factors influencing the magnitude of the apparent attraction are merged into the gravitational constant, G_1. Distance between the disturbances is the average distance between all points on the perimeters of the disturbances, facing each other. For explanations in this book, distance between the nearest points on their perimeters may be considered as the distance between the disturbances. Distance between disturbances modifies the magnitude of apparent gravitational attraction between them, in two ways.

In the first case, it alters the extent of 2D energy field in between the disturbances and causes a variation in the resultant magnitude of gravitational actions on each of the disturbances (as explained above). Reduction in the extent of 2D energy field reduces the gravitational action from the space between the disturbances. Gravitational actions from the spaces on the outer sides of the disturbances remain constant. Resultant of the gravitational actions from the outer side and from in between the disturbances, on each of the disturbances creates inertial actions of the disturbances towards each other. Apparent attraction between the disturbances is increased by a reduction in the distance between them.

In the second case, a variation in the distance between the disturbances, vary the angles subtended by them on each other. A change in the angle subtended produces a variation in the extent of 2D energy field, acting on each of the disturbances, in the direction of the other. This changes the magnitudes of gravitational actions on both the disturbances. As a result, magnitude of apparent gravitational attraction between the disturbances changes.

In figure 3F, let 'A' and 'B' be two disturbances in a 2D energy field plane. Extent of 2D energy field, acting on both disturbances from the space between them, is DEIH. This is much less than the extent of 2D energy field acting from outer sides of these disturbances. 2D energy field of extent within the sector CDEF acts on disturbance 'A' and 2D energy field of extent within the sector GHIJ acts on disturbance 'B'. As the distance between the disturbances is reduced, angle subtended by the disturbances on each other increases. Higher subtended angle, in turn, increases the area of 2D energy field, applying gravitational force/pressure

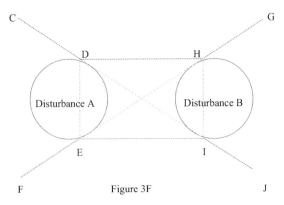

Figure 3F

on the disturbances from outer sides, while the sweep of 2D energy field in between the disturbances, more or less, remains the same. Only its extent is varied due to the change in the distance between the disturbances.

Let these disturbances be of same size, as shown in the figure 3F. Then the sweep of the 2D energy field in between the disturbances remains constant for any distance between them. Angle subtended on each other varies as the distance between them is varied. Change in the angle subtended produces a difference in the extent of 2D energy field in contact with the perimeters of the disturbances. Changes of extents of 2D energy field vary the magnitude of gravitational actions on each of them, in the direction of the other. Change in the magnitude of gravitational action, on each of the disturbances, contributes to the variation in the apparent gravitational attraction between the disturbances. Such variation is in addition to the variation caused by the extent of 2D energy field on either side of the disturbances, due to change in the distance between them. In case of 3D matter-bodies, all basic 3D matter particles are of the same (radial) size and very small. Hence, variation of the apparent attraction due to gravitation by a change in angles subtended by the disturbances on each other is appreciable only at extremely small distance, between the basic 3D matter particles of the bodies. In the present-day equation for the apparent attraction due to gravitation, this is not taken into consideration. It is one of the reasons, why the equation breaks down for very small distances between micro bodies.

When apparent gravitational attraction between more than two disturbances is considered, angle subtended makes considerable difference in the sum of magnitude of apparent attraction due to gravitation between them. E.g: sum of apparent gravitational attractions between a reference-disturbance and a twin-disturbance increase as the constituents of the twin-disturbance are moved across the line of sight from the reference-disturbance. If the twin-disturbance is a united entity, apparent attraction between the reference-disturbance and the twin-disturbance vary as the distance between constituents of the twin-disturbance changes, without varying the distance between the reference-disturbance and the twin-disturbance. An increase in the distance between the constituents of the twin-disturbance increases the sum total of apparent gravitational attraction between the reference-disturbance and the twin-disturbance. This increment is the result of increase in the gravitational action on the main disturbance, caused by an increase in the angle subtended.

Figure 3G shows a main-disturbance 'A' and a twin-disturbance constituted by disturbances 'B' and 'C'. Distance between disturbance 'A' and the twin-disturbance is very large compared to the distance between disturbances 'B' and 'C'. Extent of 2D energy field applying the gravitational efforts, contributing to the apparent gravitational attraction between disturbance 'A' and the twin-body, are: FBG on the disturbance 'B', HCJ on the disturbance 'C' and DAE on the disturbance 'A'.

As long as the distance between disturbance 'A' and the line

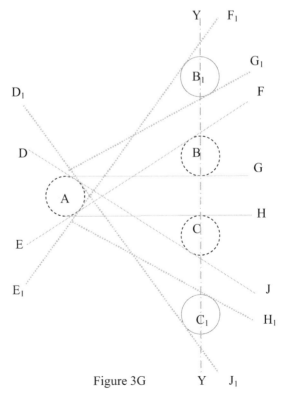

Figure 3G

YY (joining disturbances 'B' and 'C') does not vary, the angle subtended by each of the constituent disturbances 'B' or 'C', independently, on the main-disturbance 'A' remains constant. If disturbances 'B' and 'C' of the twin-disturbance move apart to the positions shown as B_1 and C_1, extent of 2D energy field applying gravitational actions on them, contributing to the apparent gravitational attraction, remains constant. Angle $F_1B_1G_1$ is equal to angle FBG and angle $H_1C_1J_1$ is equal to angle HCJ. However, the extent of 2D energy field, acting gravitationally on disturbance 'A', changes to be D_1AE_1. Contribution of the main-disturbance towards the apparent gravitational attraction increases considerably, without any variations in the parameters of the disturbances or a change in the distance between the main-disturbance and the twin-disturbance.

This phenomenon is applicable in case of composite 3D macro bodies also. Magnitude of apparent gravitational attraction between a main-body and a twin-body (composite 3D macro body) increases as the distance (perpendicular to the line joining the main-body and the twin-body) between the constituents of the twin-body is increased, within reasonable limit. High eccentricity of very large oblong spherical bodies may produce similar results in its apparent gravitational attraction, depending on its relative position, with other bodies. Changes in the magnitude of gravitational attraction are the contribution of the main body, due to the variation of extent of 2D energy field, subtended on it.

3.4.3. Many bodies in the same plane:

Apparent attraction due to gravitation between two bodies is caused by the resultant of gravitational actions on them. Hence, it may be determined only between two bodies. More than two bodies cannot have a single value of apparent attraction between them. Each of the bodies in a group may have different (or same) magnitudes of apparent attraction due to gravitation in conjunction with any other member of the group. For this, effect of presence of other bodies in any relative positions with the bodies considered, is very little. Even macro bodies cannot have apparent attraction due to gravitation in groups or in parts. Groups of independent bodies, considered as single body during calculation of apparent attraction due to gravity towards an external body, may introduce anomalies in the calculation as given in previous paragraph.

Consider three 2D disturbances in a straight line, as shown in figure 3H, whose planes coincide. Any two disturbances may be considered to form a pair. Since all of them intercept same 2D energy fields, apparent gravitational attraction will be effective between each pair. In figure 3H, disturbances 'A', 'B' and 'C' form three separate pairs; 'A &C', 'A & B' and 'B & C'. Distance between disturbances 'A' and 'B' and the distance between disturbances 'B' and 'C' are equal. We shall consider gravitational actions on these disturbances (with respect to disturbance 'A'), in the direction parallel to the line joining them. Small dashed arrows show the magnitude of gravitational action on each disturbance and black arrows show the resultant action on each disturbance.

Consider gravitational actions on the pair formed by disturbances 'A' and 'C'. Gravitational actions, 'f_1 and f_2', on both the disturbances give resultants, F_1, as apparent attraction towards each other. Gravitational actions, f_2, from either sides of disturbance 'B' or their resultant (being zero) does not affect the apparent attraction between disturbances 'A' and 'C'.

Consider gravitational actions on the pair formed by disturbances 'A' and 'B'. Gravitational actions, f_1 and f_2, on disturbance 'A' gives a resultant gravitational action, F_1, on disturbance 'A', as shown by the black arrow. Extent of 2D energy

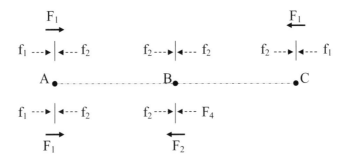

Figure 3H

fields about disturbance 'B' in the direction of disturbance 'C' is limited up to disturbance 'C'. However, gravitational action available at disturbance 'C' is 'F_1' towards disturbance 'B'. Gravitational action 'f_2', by the extent of 2D energy fields up to disturbance 'C' and the gravitational action 'F_1', available at disturbance 'C', together form a resultant 'F_4' on disturbance 'B'. Gravitational actions on either sides of disturbance 'B', (F_4 from the right and 'f_2' from the left) form a resultant, 'F_2' on disturbance 'B' towards disturbance 'A'. Apparent gravitational attraction between disturbances 'A' and 'B' is the resultant of 'F_1' from the left and 'F_2' from the right.

Similar gravitational apparent attraction is available between disturbances 'B' and 'C'. Hence, it may be noticed that presence of any number of external bodies in the same plane does not affect the magnitude of apparent attraction between two matter particles or stable macro bodies.

*** *** ** *

Chapter FOUR
CREATION OF 3D MATTER

4.1. Size reduction of a disturbance:

By its inherent nature, a 2D energy field reduces the size of a disturbance in it, to a minimum. To achieve this, the 2D energy field applies inward-acting gravitational pressure all around the disturbance in its plane. Normally, there are very few quanta of matter in a gap, formed in the 2D energy field compared to the space available and they are independent of each other. Density of (quanta of) matter within the gap is very low. Because of the expanding nature of quanta of matter and the apparent repulsion between their bodies (when their ends are in contact), these quanta of matter tend to spread out in an effort to re-form and become part of the 2D energy field latticework structure, to fill up gap in it. The process of re-formation of latticework takes some time. Before this can take place, the 2D energy field closes-in from all around, to reduce the space available (area) in the gap.

Gravitational pressure, applied by the 2D energy field, gathers loose quanta of matter (within the gap) together and opposes their expanding tendency by compressing them to form a disturbance. Once the size reduction starts, due to lack of space, the quanta of matter within the gap are unable to re-form and become part of the 2D energy field. Gravitational pressure compresses the disturbance to reduce its size. 2D energy field acts like a stretched skin around the disturbance. Thus, it provides an adhesive force around the disturbance, akin to the 'surface tension' in a liquid drop. As the quanta of matter in the disturbance are moved nearer to each other, internal pressure of the disturbance increases, until it reaches highest value when all the quanta of matter are in contact with each other, without gaps between them. Gravitational pressure, acting on the disturbance, from all around makes it possible for the disturbance to exist as a single composite body against the spreading-out tendency of its constituent quanta of matter. Surrounding gravitational pressure maintains the integrity of the disturbance.

When quanta of matter are free, within the gap in a 2D energy field, they tend to increase their lengths. Distance between adjacent junction points in the 2D energy field, trying to hold a free quantum of matter within a gap in it, are less than the length of a free quantum of matter within the gap. As the disturbance is reduced in size, sizes of the free quanta of matter within the gap also reduce. Reduction in the size of the gap compels the 2D energy field latticework to collapse around the shrinking disturbance. As the latticework collapses, in proportion to the reduction in the size of the disturbance, distance between adjacent junction points of the 2D energy field (at the perimeter of the disturbance) is always maintained much smaller than the length of a quantum of matter in the disturbance. Junction points of 2D energy fields, surrounding the disturbance, are as close as required to prevent free quanta of matter escaping from the disturbance. Any quantum of matter that escapes the junction points, ultimately makes contact with a quantum of matter of the 2D energy field latticework. Both of them, being in the same plane cannot co-exist. Hence, the quantum of matter in the 2D energy field latticework prevents further extension of the quantum of matter in the disturbance.

Due to very small size of a disturbance, its perimeter has very high convex curvature. To attain balance of forces at its perimeter, the disturbance has to have higher internal pressure compared to the gravitational pressure acting externally on its periphery. Pressure difference, needed, increases as the size of the disturbance reduces. Therefore, the rate of reduction of the size of a disturbance increases as its size becomes smaller, until it is counterbalanced by the jamming effect of the surrounding 2D energy field latticework squares.

4.1.1. Contraction of a small disturbance:

As a 2D energy field closes-in on a gap in itself, liberated and loosely scattered quanta of matter in the gap are collected together. Gravitational pressure continues to act on this collection of quanta of matter against their tendency to spread out and expand. As the quanta of matter within the gap are moved nearer, internal pressure in the disturbance increases. Perimeter of the disturbance becomes smaller and thus reducing its size. If the disturbance is too small (having very few quanta of matter in it), it may not be able to provide enough internal pressure to balance the external pressure due to gravitation, at any stage of its compression.

In normal cases, a deformation in the 2D energy field latticework spreads out and it is absorbed into the latticework. When the gap in 2D energy field, containing a disturbance is very small, latticework squares closing-in onto the disturbance jam together (adjacent junction points approach each other too close) to form gravitational field about the disturbance as shown in figure 4A.

Figure 4B shows the representation of the deformation in a larger area of the gravitational field in the 2D energy field around the disturbance. Reaction from the latticework structure prevents the quanta of matter in the 2D energy field from coming too close to each other. Certain work is stored in the quanta of matter in the form of changes in their dimensions. In this condition, the deformation in the latticework structure can neither spread out

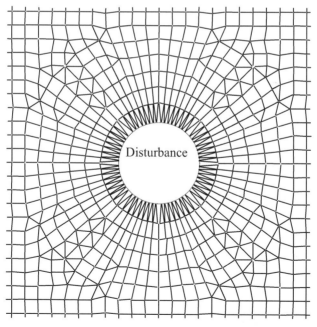

Figure 4A

nor is the latticework structure able to reduce of the size of the gap any further. 2D energy field-latticework structure reaches an equilibrium state and the 2D energy field-latticework squares will continue to stay in that jammed state around the disturbance. Most of the latticework squares, around the disturbance change their shapes to become triangles with their apex in contact with the disturbance. Matter density within the disturbance remains at a lower level than the highest value of matter density in 2D space system.

As the gap, become smaller in radial size, distortions in the gravitational field (2D energy field) increase, as shown in the figures 4A and 4B. Increase in the distortions produces higher strain in the latticework. Further distortions into the latticework structure can be introduced only by overcoming the stress, present

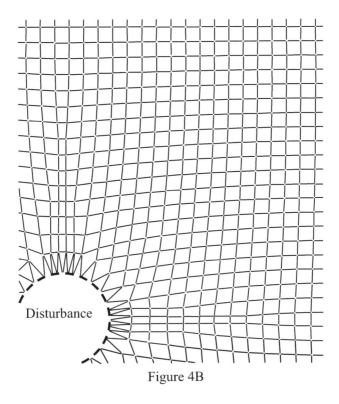

Figure 4B

in the latticework structure. As the contraction of the disturbance proceeds and the gap in 2D energy field become smaller, gravitational pressure has to act against an increasing stress in the latticework structure. Hence, the effectiveness of gravitational pressure on the disturbance reduces as the radial size of the disturbance becomes smaller. Internal pressure of the disturbance, required to balance the external gravitational pressure, diminishes.

At the same time, as the size of a disturbance becomes smaller, excess internal pressure required to balance the external gravitational pressure at its periphery, due to the increase in convex curvature, increases. Since the quanta of matter, contained in this disturbance, are few and the 2D energy field cannot reduce the gap any further, the disturbance within the gap will continue to remain as a low-matter-density 2D disturbance. This equilibrium state is the result of inability of the 2D energy field to close-in any further, rather than a balance between internal and external pressures about the disturbance.

Matter density of a quantum of matter is constant. Matter density of a 2D disturbance is its total matter content divided by its area. Since the disturbance is not contracted fully, there are gaps between the quanta of matter in it. This reduces the matter density of the 2D disturbance to lesser value. Such unsaturated 2D disturbances remain as 2D objects in 2D energy fields. They may merge with other disturbances in the same plane under the action of apparent gravitational attraction (between them). Unsaturated 2D disturbances are circular in shape and may develop linear motion within their plane under the action of external forces.

4.1.2. Contraction of larger disturbance:

A larger disturbance (with more quanta of matter in it), under contraction, has a different destiny compared to a small disturbance, as described in the last paragraph. As the size reduction of a disturbance continues under gravitational pressure, scattered quanta in the 2D energy field-gap are collected together and compressed. As the quanta of matter in the disturbance are moved nearer to each other, matter density within the 2D energy field-gap, increases. In order to create internal pressure in a 2D disturbance, reaction (resistance) to constriction forces of the quanta of matter are invoked. Resistance offered by a disturbance to the external gravitational compression is experienced as an internal pressure within the gap in 2D energy field. Development of internal pressure, in a disturbance under contraction, is carried out in two stages.

4.1.3. Internal pressure of a disturbance:

Quanta of matter in a disturbance have no particular arrangement. As they are collected together, they are oriented at random. Most of the quanta of matter in the disturbance are compressed along their axes from their ends in their own dimension. Compression along their axes reduces length and increases width against their natural tendency to elongate. Enlargement in width increases the curvature of quantum of matter's sides and reduces curvature of their end-points. Self-constriction and self-elongation tendencies within the quantum of matter are reduced. This is the first stage in the development of internal pressure and it lasts until neighboring quanta of matter in the disturbance start to make bodily contact with each other, in the same plane.

Due to the growth of quanta of matter in the disturbance in width, gaps between them reduce in sizes. Adjacent quanta of matter make bodily contact in the same spatial dimensional plane. Bodily contacts restrict their growth in that direction. As the matter density within the disturbance increases, each quantum of matter in the disturbance tends to press against its neighbors. Peripheries of quanta of matter lose their curvature and tend to become straight. Reduced curvature of perimeter segments of a quantum of matter increases self-constriction-forces in it. Each quantum of matter, making bodily contacts with more than one quantum of matter, assumes a shape that has more than two straight sides and as many end points. Straight sides of a quantum of matter cause very high self-constriction-forces. End-points gradually convert to corners. Curvature of a corner is very high. Due to high curvature of end-points, self-elongating-forces at the corners become very high. Matter content in each quantum of matter tends to expand into any gap available, between them. This tendency, to grow outwards at the corners, produces reactive efforts against (resistance to) the external gravitational pressure and appears as the increasing internal pressure of the disturbance. In this second stage, the external gravitational pressure acts against increasing resistance offered to further compress the disturbance.

Due to the compressive efforts (in their own spatial dimensions) from their ends, quanta of matter in the gap are reduced to their minimum magnitude in the first-dimensional space system (length). Further compression compels them to grow into the second-dimensional space system. Thus, each of the quanta of matter becomes a two-dimensional body. Its length is reduced, with corresponding increase in its second-dimension – the breadth. When all the quanta of matter in the disturbance have grown fully into the second-dimension, this closely packed body of two-dimensional quanta of matter makes the disturbance a two-dimensional circular sheet of pure matter, in a plane.

At the same time, each quantum of matter in this two-dimensional body (disturbance) keeps its separate identity, by each one becoming an independent (probably, hexagonal) 2D body. Their body shapes are such that they fit close with each other without gaps between them. They continuously strive to attain their stable free state in the single-dimensional space system by reducing in width and expanding or spreading-out in length. This tendency of quanta of matter produces a resistance against the gravitational pressure acting on the disturbance from all around the perimeter of its matter-body. Gravitational pressure is the root cause for the production of this resistance by the disturbance. Resistance against the gravitational pressure, produced by the disturbance, gives rise to its 'internal pressure'. Internal pressure of a disturbance does not increase the matter density of the quanta of matter in the disturbance. Attempt to increase the internal pressure compels the quanta of matter to grow into higher dimensional space systems. Gaps in the disturbance are filled up first. After the disturbance has been fully converted into two-dimensional state, further attempt to increase the internal pressure compel the quanta to grow into 3D space system.

Horde of free quanta of matter, within the disturbance, bestows the disturbance with its matter content. Matter content of a disturbance is, more or less, distributed evenly within the area of

Chapter 4. CREATION OF 3D MATTER

the 2D disturbance. In 2D space system, relation between the matter content of a disturbance and its area is its matter density. During the initial stage of compression, quanta of matter in the disturbance are not closely packed. There are gaps between adjacent quanta of matter. [Gaps between quanta of matter should be understood with respect to the spatial dimension of the disturbance. This space is filled with quanta of matter in 2D energy fields in other planes.] Matter density of the disturbance increases as adjacent quanta of matter in it are moved nearer (during the contraction of the disturbance) until all the quanta of matter, in the spatial plane of the disturbance, are closely packed. At this stage, the matter density of the 2D disturbance is equal to the matter density of a quantum of matter, which is the highest density of matter in nature.

A 2D disturbance may reach its stable state only when its internal pressure, developed due to the compression of its constituent quanta of matter, balances the external gravitational pressure applied on it. Due to the convex curvature of disturbance's perimeter, in its stable state, magnitude of internal and external pressures, required for a balanced state, are not always equal. Internal pressure required is always greater than the external pressure. (Similar to internal pressure of a liquid droplet and external pressure applied on it). Difference between them, required for a stable state, increases as the diameter of the circular 2D disturbance shrinks.

4.1.4. Very large disturbance:

[Adjectives denoting sizes of disturbances are comparative to the radial size of a corpuscle of light]. A very large disturbance, in a plane, being a big circular body, has low convex curvature at its perimeter. Difference between the internal pressure of the 2D disturbance and the external gravitational pressure on it, required to arrive at a balanced state at its perimeter, is relatively small. Therefore, the internal and external pressures of a large 2D disturbance are nearly equal and difference between them is very little. Gravitational pressure, acting on the disturbance, may not be able to compress the disturbance or the compression may be too slow to make appreciable changes in the state of the disturbance in short time. Many of the quanta of matter, in such a disturbance will gradually regain their freedom and become part of 2D energy fields until the size of the disturbance is greatly reduced.

Quanta of matter from the peripheral area, one by one, will find suitable places in the surrounding 2D energy field and become part of its quanta-chains in the latticework structure. This process will continue until the size of the 2D disturbance has come down, sufficiently, so that its peripheral curvature is of the required level to commence further contraction of the 2D disturbance. General nature of the 2D energy field also affects the limits for the contracting stages. Inability of the 2D energy field to continue the development of very large 2D disturbances prevents the natural formations of very large (by matter content) 'basic 3D matter particles' in nature.

4.1.5. Disturbance of optimum size:

As the size of a very large disturbance is reduced, gradually, curvature of its periphery increases. Most of its quanta of matter are in their advanced stage of full-conversion into 2D state. Conversion of quanta of matter into second spatial dimension increases the curvature of their sides and reduces the internal pressure of the disturbance. 2D disturbance is able to contract at a faster rate. As the size of the 2D disturbance is reduced further, jamming effect on the 2D energy field (as explained in the case of very small disturbances. Refer Section 4.1.1) also comes into effect. As the bodies of neighboring quanta of matter in the same plane make contact with each other, their sides are straightened. This increases the self-constriction effects and reduces self-elongation effect in the quanta of matter. Both these effects tend to move the matter contents of quanta to fill up the gaps in the 2D disturbance (in its spatial plane) and thereby increase its matter density and internal pressure. Growing internal pressure and waning external pressure (due to jamming effect of 2D energy fields) limits the size reduction of the disturbance, when they are in balance. As and when the size of the disturbance reaches an optimum magnitude (or if the disturbance is of smaller size, but not too small), the curvature of its perimeter may not be large enough to provide the required pressure difference for further compression.

Due to the higher curvature of the 2D disturbance's perimeter, its internal pressure, required to balance the external pressure (applied by the 2D energy field), remains higher until the disturbance is reduced to an optimum size. At the optimum size, these pressures balance and the contraction of the disturbance stops. In this state, the disturbance is a 2D matter-body of maximum matter density. There are no gaps between constituent quanta in it. Gravitational pressure around the matter-body of the 2D disturbance produces an effect similar to the surface tension around a liquid drop. 2D disturbance behaves like a liquid drop with respect to its internal and external pressures. If there is just sufficient matter content in a 2D disturbance to produce a 2D matter-body of maximum matter density, when its internal pressure is balanced by the external pressure on it, it may be said to have a radial 'critical size'. At its critical radial size, matter density of the disturbance is equal to the matter density of a quantum of matter. Internal pressure balances the external gravitational pressure on it. All quanta of matter in a critical-sized 2D disturbance are converted to their 2D states and there are no gaps in the disturbance. Thus, this disturbance becomes a saturated 2D matter particle. A saturated 2D disturbance/matter particle is circular in shape. It exists in the plane of a 2D energy field. It has maximum possible matter density in the 2D space system (which is the same as matter density of a quantum of matter).

4.2. Creation of three-dimensional matter:

If the size of a saturated 2D disturbance is larger than the critical radial size of a 2D matter particle, it has comparatively higher external gravitational pressure on it and lower internal pressure from within. Hence, this body may be further compressed due to the difference in its internal pressure and the external pressure on it. Magnitude of a 2D disturbance is the length of its perimeter. In order to reduce its magnitude, the 2D disturbance has to reduce to still smaller size in its plane. However, all constituent quanta of matter of the disturbance have converted to their 2D state to fill up the entire space within the perimeter of the disturbance. Matter content in the disturbance has already reached the minimum possible size in the two-dimensional space system.

Since the matter density within the 2D disturbance has reached its highest value, the 2D disturbance cannot be contracted any more in 2D space system, unless part of its matter content is removed from the plane of the disturbance. Constituent quanta of matter of the saturated 2D disturbance, growing into the third dimensional space (developing thickness), can remove part of their matter content from the 2D space system.

Further contraction of the saturated 2D disturbance can take place, only when its constituent quanta of matter grow into the third spatial dimension to pave way for a reduction of the area of the disturbance in its 2D plane. That is, under further application of higher external gravitational pressure, constituent quanta of matter and the saturated 2D disturbance constituted by them will now grow in thickness. Consequently, this 2D disturbance is converted into a 3D disturbance – a three-dimensional matter particle. In our (3D world) sense, it is now, that a 'real matter particle' (3D matter) is created from the 'functional matter particles'. This matter particle has become fit to be considered as 'real' due to our ability to observe and measure its parameters by using the standards of 3D space system. Until it was converted into a 3D matter particle, all its parameters were intangible to us. Reality of the matter particle has not changed but the matter particle has now become observable by us, the rational beings. The particle and its constituent quanta of matter, though real, they were unobservable by us during their 1D and 2D states. A 3D matter particle is a disturbance with respect to all 2D energy fields in the planes passing through it. Hence, it may be called as a '3D disturbance' or as a 3D object.

Once a disturbance has developed into the third spatial dimension, it has certain thickness and the disturbance simultaneously exists in more than one 2D energy fields. When the disturbance is a 2D body, it exists in its own plane and co-exists with 2D energy fields in all other planes, which are at an angle to its disc plane. In its 3D state, the disturbance is not only a part of more than one 2D energy fields but it breaks the continuity of all the 2D energy fields, in which it exists. All quanta of matter in their functional state (constituents of 2D energy fields), which happen to be at the place of formation of the 3D matter, withdraw themselves out of the space where the 3D matter is being formed. Since the 2D energy fields can act (mainly) only on the curved surface of a disturbance, most (gravitational) interactions between the 2D energy fields and the 3D disturbance are limited to the circumference of the 3D disturbance, where its curved perimeter is situated.

As soon as a 2D disturbance is converted into a 3D object, by growing in thickness, continuity of all 2D energy fields in the planes passing through the 3D object are broken. In this condition, thickness of the newly formed 3D object is more than the distance between two adjacent parallel 2D energy fields. Growth in thickness facilitates further reduction of the disturbance's area in its original 2D plane. Matter content of the disturbance, which was limited to its area in the 2D plane is now distributed into a volumetric space in three-dimensional space system. Due to their growth into the third dimensional space system, sizes of the quanta of matter (within the disturbance), in first and second spatial dimensions, can now be further reduced. Their areas are reduced and their thicknesses (into the third dimensional space system) are increased.

Gravitational pressure is now applied on the disturbance by all the 2D energy fields in which the 3D disturbance exists, each one in its own plane. However, their actions are limited mainly to the planes, where the 3D disturbance has curved perimeter. As the contraction of the disturbance proceeds, matter content of the 3D disturbance continues to grow in thickness into the adjacent 2D energy fields. Each of the quanta of matter, in the 3D disturbance, keeps its individuality and the disturbance grows in thickness towards becoming a (spherical) body of pure matter. In their closely packed state, shapes of the quanta of matter become more like cubes or other solid shapes than spheres. Not all the quanta of matter in a disturbance transform simultaneously. They change at random, depending on their orientation and the direction of compressive pressure on them. They also may rearrange themselves to redistribute evenly within the 3D disturbance. The process of transformation continues until the radial size of the disturbance (in the main disc planes of its existence) reaches the critical radial size. At critical radial size, internal and external pressures at the curved periphery of the 3D disturbance, in each plane, balance each other.

Similar processes take place, simultaneously in all planes within the gap in the 2D energy fields. Part-disturbances in all these planes combine, jointly to form a single 3D object. If the quanta of matter in the 3D disturbance are in sufficiently large numbers, (this is a hypothetical case) this process will continue until the disturbance achieves a spherical shape of critical radial size, circular in each of the planes passing through the disturbance. In balanced condition of conversion, the disturbance may continue to stay in its present status as a 3D disturbance until it is assimilated into another 3D disturbance under action of apparent gravitational attraction.

4.2.1. Creation of higher-dimensional matter:

Should the volume (radial size) of the (hypothetical) 3D disturbance-sphere be more than the critical radial size, its surface curvature may produce enough pressure difference between internal and external pressures to continue its compression by the gravitational pressure. Further compression of the disturbance is possible only if the volume of the disturbance-sphere, which is already at its highest matter density, can be reduced. Reduction in the volume of the 3D disturbance is possible only if part of its matter content can be removed from the 3D space system. If the resultant pressure on the disturbance is sufficient, quanta of matter in the 3D disturbance start to grow into the (imaginary) fourth spatial dimension. Growth of the disturbance into the fourth spatial dimension will allow a reduction in its 3D volume until internal and external pressures can be balanced. The 3D disturbance may continue its growth into the fourth spatial dimension until its radial size in 3D space system reaches critical diameter.

We may assume that if the matter content of the disturbance is sufficiently large, similar process may repeat for the conversion of the matter particle into higher-dimensional space systems. Since we know nothing about the fourth and higher-dimensional space systems this is only a hypothetical case in the development of matter. We cannot venture to hypothesis on the properties of matter in its 4D spatial state. [The 'time', which is presently regarded by many people as the fourth dimension, is not a dimension of the space. We may realize a fourth spatial dimension only when we are able to partition the space by four mutually perpendicular planes.]

4.2.2. Critical radial size of a 3D disturbance:

Radial size of a stable 3D disturbance, in each of its (disc) planes, is of critical value. At its critical radial size, internal pressure produced by the (highest possible) matter density of the disturbance and the external pressure applied on the disturbance, due to the gravitation by each of the 2D energy fields in its own plane, balance each other. As the radial size of a disturbance decreases, in any plane, its internal pressure increases due to the compression of its matter content against the force of self-constriction of the constituent quanta of matter. At the same time, external pressure applied by the 2D energy field (gravitational pressure), diminish due to the jamming effect of the latticework squares. These two effects balance at the critical radial size of a 3D disturbance, in any plane.

At critical radial size, matter density of all 3D disturbances in nature is the same and it is of constant value. This is the highest matter density for 3D matter in the universe and it is equal to the matter density of a quantum of matter. Should the density of matter tends to increase beyond this value, due to larger radial size of the disturbance and it has already occupied full volume in 3D space (it has become a sphere of greater diameter than critical radial size), matter content of the 3D disturbance will start growing/ converting itself into an object in the four-dimensional space system.

We have considered the gravitational pressure, acting only on the curved perimeter of a contracting disturbance. In each of the planes of 3D disturbance's existence, 2D energy field of that plane acts on its curved perimeter. Gravitational actions by the 2D energy fields on flat surfaces of a 3D disturbance are negligible. Applications of gravitational efforts on all surfaces are the same but its actions are limited to the curved surfaces only. (Refer section

3.3.4). A flat surface on a 3D disturbance is made up of straight-line perimeters in each of the 2D planes passing through the disturbance. These planes are at oblique angles to the mean disc plane of the disturbance.

Gravitational actions on the curved surface (circumferential perimeter) of a 3D disturbance in each of the 2D energy fields are inward (towards the centre of the disturbance) to compress the disturbance. Faces of the 3D disturbance, being (almost) flat, are not affected by the gravitational actions on them by the 2D energy fields, which are at an angle to the median plane of the disturbance-disc. Gravitational action on its faces, balances the internal pressure on the flat surfaces of the faces of the 3D disturbance. If these disc-faces are flat, there is no resultant effort acting at the flat surfaces of a 3D disturbance, to produce their motion. Only a variation of disturbance's internal pressure can produce appropriate resultant action at these flat surfaces and cause a change in the curvature of surface or produce a linear motion of the surface.

4.2.3. Shaping up of a 3D disturbance:

Assuming that there is no other external efforts on a 2D disturbance, and it has enough quanta of matter, its contracting process will continue until all quanta of matter in the disturbance are converted into their 3D states. Quanta of matter in the plane of 2D disturbance have grown into the third spatial dimension to gain thickness. All the quanta in the disturbance are now shaped like cubes or other volumetric shapes that fit close with each other, rather than circular. The disturbance, as a whole, has grown in thickness to become a disc-like 3D object.

Corners of this disc, where its faces meet the cylindrical surface of the disc have very high curvature (because these surfaces are at right angle to each other). 2D energy fields in planes other than the disc planes of the disturbance, now act on these corners. Gravitational action on them, by the 2D energy fields, tends to flatten these corners. This process compels the disturbance to reshape itself to become a spherical body. If enough time and quanta of matter were available for the process, by the time all quanta of a disturbance have converted to become 3D matter; the disturbance would have assumed spherical shape. The disturbance would have become a sphere of pure 3D matter.

Growth of a disturbance into higher dimensional space than the third dimensional space will take place if a very large number of free quanta of matter (or exceptionally large quanta of matter) are available in the disturbance and enough time is available before any non-isotropic condition develops within the disturbance. Under such condition, a disturbance will grow into a higher-dimensional matter-body. Since, at present, this is only a hypothetical case we will not pursue it. Normally, availability of quanta of matter in a disturbance is limited, that it will not be able to grow into even a very small part of a sphere. In this case, the disturbance grows into a disc of 3D matter until other external actions influence it. Development of high matter content-3D disturbances may take place by amalgamation of more than one 2D or 3D disturbances (or quanta of matter of exceptionally large matter content).

4.3. Development of 3D matter particle:

It may be noted that the matter content of a disturbance is quantified in terms of number of quanta of matter. Matter content of a 2D or 3D body may be varied only in terms of integral numbers of quanta of matter. Due to variations in the matter content of different quanta of matter, matter content of a 3D matter particle may not always correspond to the number of quanta of matter in it. Nevertheless, the 'real' matter content of the disturbance (3D body) changes gradually due to gradual conversion of functional quanta of matter into 'real' matter. At the end of the conversion of a disturbance, all the quanta of matter (it contains) will be 3D matter particles and the disturbance will be of highest 3D matter density. Matter density of all 3D matter-bodies – basic 3D matter particles - in the universe is the same as that of a quantum of matter.

Creation of 'real matter', explained above, is confined to disturbances formed only in single plane. Usually, the disturbances or gaps developed in 2D energy fields exist simultaneously in many planes. Actions of creation of 'real matter' are identical in all these planes and they take place simultaneously and in association with each other.

4.3.1. Ejection force:

During its contraction, a 2D disturbance in a 2D energy field grows into adjacent 2D energy fields and thus attains thickness to become a 3D disturbance. This 3D disturbance has two faces situated in different 2D energy field planes. Each of these faces and their perimeters, depending on their curvature, may exist in one or more 2D energy fields at the same time. During this stage of development, the disturbance is neither homogeneous nor isotropic. Lack of isotropy causes asymmetry in its internal pressure. Areas, with higher internal pressure, bulge outwards into adjacent planes. Areas, with lower internal pressure, contract inward. Thus, disc-faces of a 3D disturbance-disc, during its development, is usually uneven.

When a 2D disturbance is situated within only one 2D energy field, such asymmetry will not have appreciable effects on the disturbance other than to introduce linear motion to the disturbance, in certain cases. A 3D disturbance, existing simultaneously in many 2D energy fields, is affected by such an asymmetry in a different way. Asymmetry in the internal pressure of a disturbance causes unevenness in disturbance's thickness and in the shape of its disc-faces. Thus, depending on the internal pressure, certain areas on the faces of the disturbance-disc may bulge inward or outward. Places of such unevenness on the faces of the disturbance-disc depend on the internal arrangements of the quanta of matter within the disturbance, during their compression.

Figure 4C shows the cross section of a 3D disturbance (greatly enlarged), whose thickness is equal to the thickness of few 2D energy field-planes. Consider a part 'A' (between the dashed lines) of the disturbance across the thickness of the disc. It contains a convex bulge on one side due to higher internal pressure of matter content in that part of the disturbance. Both ends of part 'A' have different curvatures. Left-side end is flat but the right-side end has a convex curvature. Difference in curvature produces an asymmetry in the gravitational action at the end surfaces of this part of the disturbance. Left side-end being flat, it is not acted upon by the gravitational effort but the right side-end, having a convex curvature, has higher gravitational effort acting on it. Thus, part

Chapter 4. CREATION OF 3D MATTER

'A' of the disturbance has a resultant lateral moving effort on it.

During the formation of a disturbance, there may be many such efforts applied at different places on its faces. Resultant of all these efforts, acting on a disturbance (as a whole) or part of a disturbance, as and when it can overcome natural adhesiveness of the 2D energy fields (provided by external gravitational pressure), gives that part of the 3D disturbance a linear motion across the 2D energy fields of its creation. Latticework structure of 2D energy fields in the direction of motion will part, to facilitate the motion of the part of the 3D disturbance. Direction of this motion is across the disturbance disc. This is another aspect of the gravitation and may be called the 'ejection force' on the 3D disturbance.

There are great many 2D energy fields on the way of motion of the 3D disturbance-part, which is being ejected. Ejected part of the 3D disturbance does not apply any effort on the 2D energy fields to part them. An effort can be applied to a 2D energy field only in its plane. Direction of motion of the ejected part of the disturbance pass through only very few 2D energy fields. In all

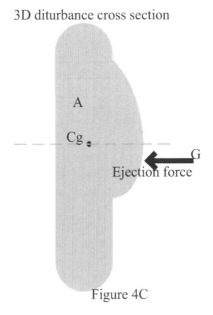

Figure 4C

other 2D energy fields, the disturbance moves across them. Resistance to its motion is experienced only from those 2D energy fields, along the planes, in which the disturbance is moving at any time. All other 2D energy fields part on their own to pave way for the ejected part of the disturbance. As soon as the 3D matter-body tends to cross a 2D energy field, few of its quanta of matter are being violated by the possible intrusion by 3D matter. At that instant, each of those quanta of matter will withdraw to either side of the 3D matter-body to produce a gap, to facilitate the existence and passage of the 3D matter-body.

Since the gap required for the passage of a 3D disturbance is, in most cases, is larger than the length of a quantum of matter, chains of quanta of matter in the 2D energy field experience axial displacements. Displacement of quanta of matter requires certain effort. The required effort is supplied by the 2D energy field to move itself out of the way of the 3D matter. This effort, supplied by the 2D energy fields, will be taken back by the same 2D energy fields during the return of their quanta-chains to their original states. Hence, there is no resultant effort or energy spent by the 2D energy fields to open and close a gap for the 3D matter's passage. As soon as the disturbance-part moves away from a 2D energy field, its quanta of matter will return to their original place in the latticework, to close the opening.

At times, radial size of the ejected part of the disturbance may be smaller than critical radial size of a 3D disturbance. In this case, gravitational actions will proceed to make changes in its shape so the ejected part of the disturbance enlarges to critical radial size.

4.3.2. Spinning force:

Asymmetry of the gravitational actions on the faces of a 3D disturbance or part of a 3D disturbance causes its ejection from and across the 2D energy fields of its development. Ejection force, itself, may not be symmetrical about the 'center (of mass)' of the ejected part of the 3D disturbance. Such asymmetry of the ejection force, about the 'center of mass' of the ejected part of the disturbance disc, produces a couple/torque about one of its diameters. This torque imparts a turning movement to the 3D disturbance-part disc. Part of the ejection force, producing the couple, is the 'spinning force' on the 3D disturbance.

Axis of spin of the ejected disturbance-disc is always at right angle to the direction of ejection. Purpose of the spinning force is only to initiate a spin motion of the ejected 3D disturbance disc about one of its diameters. Thereafter, the spin motion of the detached 3D disturbance-disc is developed and maintained at appropriate level by the natural stabilization process of the disturbance disc (discussed in the next section). Now, the disturbance-disc has two independent motions, a linear motion in space and a spin motion about one of its diameter.

In the natural development of a disturbance, its growth into the third dimensional space takes place simultaneously in many planes, in the 3D space. Ejection and spinning forces on the disturbance-part may not develop in all the planes, simultaneously or they may not be of equal magnitude. Spinning force on a 3D disturbance is another aspect of the gravitation.

4.3.3. Ejection and spin of a disturbance:

During the contraction of a 3D disturbance, due to the asymmetry in its matter content distribution, ejection forces develop on its disc faces. Ejection force, as and when it is able to overcome the adhesiveness of matter content within the disturbance and the resistance to its linear motion from the surrounding 2D energy fields, pushes-out a part (or whole) of the disturbance around the point of application of the ejection force. This part of the 3D disturbance is at saturated matter density level in the 3D space system. The ejected part of the 3D disturbance disc, moving out of the 2D energy fields of its creation, will soon adjust its disc size and shape by itself (with the help of gravitational actions from surrounding 2D energy fields) to stable conditions.

In case, enough ejection force does not develop on its disc face, the 3D disturbance will go on reducing its radial size to the 'critical disc size' while growing into the third spatial dimension and become a sphere of matter content (real 3D matter). Critical disc size is the radial size of a 3D disturbance, at which its internal pressure from the compressed quanta of matter and the external compressive pressure applied by the gravitational action balance each other. It is the delay in developing the required ejection force, during the contraction period of the disturbance, which contributes towards the ejected disturbance becoming thicker. More delay in the ejection makes the ejected part of a disturbance thicker and hence the ejected 3D disturbance will have higher matter content.

There is no definite mechanism in a 3D disturbance to control or limit the development of sufficient ejection force. Consequently, magnitude of matter content or the radial size of any ejected part of 3D disturbance are arbitrary. However, the status of the surrounding 2D energy fields and the rate of availability of free quanta of matter, during the formation of the disturbances, do have certain effects on the development of the ejection force. Hence, the majority of the ejected 3D disturbance-parts in a particular external environment may be of similar (average) matter content level. A change in the external environment, normally, changes the matter content pattern of the ejected disturbance-parts. This also ensures that other than in specially conceived environment, ejected 3D disturbance-parts may include 3D disturbances of different magnitudes of matter contents.

Ejection force imparts a linear motion to the ejected part of a 3D disturbance (initially) at right angle to its median plane. Since all such disturbances are very thin, median plane of a disturbance, referred here, generally means the 2D energy field plane passing through the center of disturbance and median to the planes of its faces. Spinning force initiates a spin motion about one of its diameters in the median plane of the ejected part of the disturbance. Spin motion gives mechanical stability to the ejected part of the 3D disturbance, in space. Gravitational pressure, around the circular periphery of the ejected part of the 3D disturbance, maintains the disc's circular shape and maintains the radial measurements to their critical values.

Gravitational pressures, on the faces of the ejected part of the 3D disturbance (these faces being flat surfaces), comes into effect only if the matter density of the ejected part of the 3D disturbance decreases from its critical value for a 3D matter particle. Should the matter density of the ejected part of the 3D disturbance diminish, gravitational pressure from all around its periphery will act on the ejected part of the disturbance to compress it so that to develop its internal pressure and matter density to the critical values. By doing so, radial size of the ejected 3D disturbance is restored to critical value.

4.3.4. Centrifugal force in a disturbance:

As the ejected part of the 3D disturbance (disc) starts to spin and move away from the parent 3D disturbance or the region of its development, the spinning part of the 3D disturbance disc becomes a separate individual entity, in its own right. Once, the disc develops spin motion, the original ejection force is no more perpendicular to its face or in the direction of its linear motion. Since the ejection force is produced by the 2D energy fields, its direction does not change. It is the ejected disturbance, which turns with respect to the direction of ejection force. The ejection force can be considered to be composed of two components. Whenever the disturbance disc is at an angle to the line of linear motion, one component of the ejection force acts in the direction of the linear motion, across the disc. Other component of the ejection force acts along the diameter of the disc. Due to the spin motion of the disc, components of the ejection force, along the diameter of the disturbance-disc, tends to push matter content of the ejected 3D disturbance disc towards its forward edge.

During every spin of the 3D disturbance disc, its forward and rearward edges are reversed. Thus, the matter content of the ejected 3D disturbance has a tendency to concentrate towards the forward and rearward edges of the disc. This may be taken as equivalent to the (imaginary) centrifugal force of the matter content of the spinning 3D disturbance disc. Centrifugal force is an apparent force. In this case, the accumulation of its matter content towards the periphery of the disturbance's body produces the illusion of certain force acting on it, such as to move its matter content away from disc's spin axis (the center of rotation). Movement of the matter content within the disturbance is caused by the push action of the ejection force, which is a manifestation of the gravitational action.

For the following explanations, we may consider this apparent centrifugal force due to the displacement of matter content within the disturbance as a real force. This avoids frequent mentioning relative displacement of matter content within the disturbance.

4.4. Photon:

Photons are corpuscles of light and similar radiations. Detailed explanations on the creation, development, structure, properties, apparent interactions of photons and their contribution to the development of superior matter-bodies can be found in the book 'Hypothesis on MATTER'. Only those factors, which are pertaining directly to gravitation, are briefly mentioned below.

Centrifugal force, acting within the ejected disturbance-part, exerts a pressure outward and against the gravitational contracting forces on the spinning disturbance's circular periphery. Radial size of the spinning 3D disturbance disc will become stable when the forces exerted by its internal pressure and the centrifugal force of the 3D disturbance disc, together, balance the external actions on its circumferential surfaces by gravitation. Radial size of the spinning 3D disturbance, in its stable state, is critical disc size. Critical disc size is equal for all independent stable 3D disturbances. Since all basic 3D particles in nature are made up of this type of 3D disturbances, we can say that this is the critical disc size of basic 3D matter particles in the 3D space system. Basically, there are no 3D matter particles larger or smaller in radial size than (or different from) this critical disc-sized 3D disturbance. They may differ only in quantity of their matter contents and corresponding thickness at their equatorial region. All other bodies, found in the 3D space system, are union of two or more of this type of 3D disturbances.

Gravitational pressure, acting on the circular periphery of a 3D disturbance, further helps to adjust the thickness of the ejected 3D disturbance-disc in proportion to its matter content. Difference in the gravitational actions on its disc faces provide the required ejection force and spinning force for the disc's natural motions. Shapes of the disc faces are continuously varied for this purpose. Linear speed of the disturbance disc is developed into a constant speed and its spin speed is developed to be proportional to its matter content. This disc of saturated 3D matter, moving at a constant linear speed and spinning at a speed proportional to its matter content is the core body of a 'photon'. All movements of a photon are with respect to surrounding 2D energy fields.

A photon essentially has a 3D matter core in the shape of a disc, spinning about one of its diameters. Latticework distortions, in the 2D energy fields surrounding the 3D matter-body, are required to sustain the integrity of the photon in 3D space system and its continuous motions. These 2D energy field-distortions, called 'inertial pocket', move continuously in the space at critical linear speed. Motions of the associated distortions carry the photon's matter-body through the space. Hence, the distortions, associated with the matter-body of the photon are an integral part of the photon. Both, the matter-body and the associated 2D energy field-distortions, together form a 'photon'. In this text, the term 'photon' is generally used to represent the 3D matter-body of the photon and the term 'inertial pocket' is generally used to represent the distortions associated with the matter-body of the photon.

Whenever a disturbance is formed in a region of space, in the 3D space system, distortions are produced in as many 2D planes passing through that region. All the 2D energy field- distortions together form region of distortions in the 3D space system. Interaction between the disturbance and each of the 2D energy fields is as described in above paragraphs. Hence, matter conversion takes place in all the 2D energy fields passing through the region. Each of the planes, passing through the matter-body of the photon, may have varying numbers of quanta of matter in them. After their conversion to 3D disturbances about each of the planes, all the quanta of matter in these 2D energy fields, together, constitute a 3D disturbance in the 3D space system. Ejection and spinning forces, applied on the composite 3D disturbance, is the resultants of all the efforts applied in its individual planes. In these sections of the book, interactions between a 3D disturbance and one representative 2D energy field in one plane are described. Since a 3D disturbance exists in several 2D energy fields, similar actions take place in all associated 2D energy fields separately and simultaneously.

4.4.1. Shape of a photon:

Because of its rapid movements and constant shifting of its matter content within its core body, a photon cannot have a permanently fixed shape. However, Figure 4D and 4E show (approximate) instantaneous shapes of a stable photon's matter body. Figure 4D shows the isometric view. Contour lines, on the surface of the photon disc-body in figure 4D, are given only to highlight its shape. Figure 4E shows the face view, the cross sectional plan and the side view of a stable photon's matter-body. We may say that a photon generally has a segmented spherical shape. Its diameter in the median plane, about which the photon spins, is its 'axis of spin'. Ends of the axis are photon's 'poles'. Considering a photon to be a part of an imaginary sphere, the 'equator' of a photon passes through its curved surface, midway between its poles.

Each quantum of matter, in a photon, maintains its independence and separate identity. Though the quanta of matter in a photon's matter-body are packed together, they have no mutual

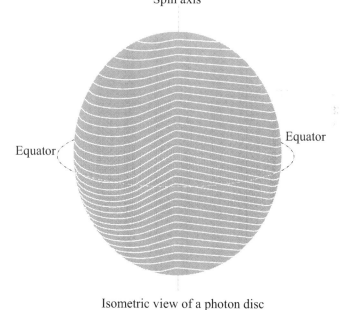

Isometric view of a photon disc

Figure 4D

affinity other than the feeble attraction between their matter contents, which are in contact with each other. In their 3D status, within a photon's core body, each quantum of matter is (somewhat) cubical or similar volumetric shape and they have no definite ends in appearance. 2D plane for actions during quantum's 2D status and the axis for its actions during its 1D status are chosen during the reversion of the quantum of matter into its free status. Shape of each quantum of matter may vary as the shape of the photon varies. At any instant, shape of photon's matter-body depends on the external pressure applied around its body.

Quanta of matter in a photon are not arranged or linked in orderly fashion. Due to their independence, quanta of matter within a photon continuously try to expand, spread-out and link together to form parts of 2D energy field latticework structure. These quanta

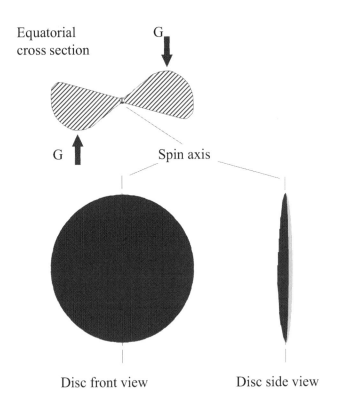

Figure 4E

of matter constitute the photon's matter content. Quanta of matter in a photon are held together by gravitational pressure acting on its body-surface. Matter-body of a photon is similar to a stretchy balloon filled with flexible marbles. Marbles represent the independent quanta in their 3D state and the gravitational pressure acting all around the photon provides the balloon. Because of this type of arrangement, matter content of a photon has a fluid behavior within its boundary. Depending on the variations in the gravitational actions, in various forms on its surface, shape of a photon's matter-body is easily changed.

As a photon builds up its linear and spinning motions, under external actions by the 2D energy fields, its matter content re-distributes and assumes appropriate shape as per physical laws. Shape, assumed by a photon's matter-body, is such that its internal and external pressures always tend to remain in balance on every point on its surface. A photon, in its stable state, is free of any adhesion (mutually interactive force) with the 2D energy fields of its existence. Should the internal or the external pressure dominate the other, interactions developed between the photon's body and the surrounding 2D energy fields, act as an adhesion between them. Region of 2D energy fields, in the immediate neighborhood of photon's matter-body, contains all necessary 2D energy field-distortions required for stable sustenance of the photon.

Photons are generally classified according to the magnitude of matter, they contain. All radiations of matter – heat, light, X-rays, gamma rays and cosmic rays – are in the form of continuous flow of the photons. Other than, for the magnitude of their matter contents, all photons are identical in shape and radial size. Variations in the magnitude of matter content produce differences in the thickness of segments of photons' matter-body. Photons may be further classified into clockwise spinning and anti-clockwise spinning photons, with respect to any external reference.

4.4.2. Concepts of a photon:

Current concept of a photon is quite different from what is explained in the preceding paragraphs. At present, a photon is considered as a quantum of electromagnetic wave/particle (?) with zero rest mass. Mass of a body depends not only on its rest mass but also on its linear speed. Hence, a photon moving at the speed of light should have infinite mass. In order to overcome this illogical result, it has become necessary to arbitrarily assume the photon to be a mass-less body.

Since a photon is presently considered to have no rest mass, its mass should always remain zero irrespective of its speed. Momentum of a body is related to its mass. Momentum is essential for a body to do work. Yet, the mass-less photons are found to do work. In order to overcome these contradictions, even though they possess no rest mass, in mathematical equations, a photon is also considered (simply) to possess certain momentum (in some cases, enough momentum to knock out orbital electrons from some atoms!). Photon is assumed to have no matter content and carries (undefined) energy, which can do work. Present theories also assume that matter is vested only in subatomic particles and larger bodies, which have rest mass. How or from where this matter came into being is not explained.

A photon, in this concept, is a 3D matter particle that has a matter content indicated by its rest mass. It also has an associated energy (as explained in the following sections), which did the work to create and sustain the matter-body of the photon or which can do external work, on the photon's disintegration. This part of associated energy is also considered as the mathematical equivalent of electromagnetic wave of the same frequency as the spin speed of the photon. Distortions, equated to the electromagnetic wave associated with a photon, exist within the 2D energy fields but coupled with matter-body of each photon. Body of the photon is made up of 'real' 3D matter. Matter-body of a photon and its associated inertial pocket (distortions) are mutually developed and maintained. Inertial pocket of a photon, at any instant, resembles instantaneous part of an electromagnetic wave. This gives a photon, its dual nature of corpuscle and electromagnetic wave. A photon is the smallest 3D matter particle that can exist independently and in reality. It is a corpuscle of light or any other similar radiation. All other superior 3D matter bodies are made up of photons.

A photon is a 3D body of (segmented-spherical) disc-shaped

3D matter content with sufficient distortions in the associated 2D energy fields to sustain its stability and integrity. Distortions in the associated 2D energy fields provide the matter-body of the photon with the required compression and motions. In this text, matter-part and the distortion-part of a photon are often treated separately. In such cases, matter-part of the photon is called the photon and the distortion-part is called the inertial pocket.

Photon moves through 2D energy fields. Moving 2D energy field-distortions (inertial pocket around the matter core of the photon) carry the matter-body of the photon. There is a relative motion between the photon and the 2D energy fields. Relative motion causes resistance to the motion of the photon. However, at any instant, sufficient ejection force is produced by the inertial pocket to overcome this resistance. Since, both the resistance and the ejection force are produced by the 2D energy fields; they effectively reduce any drag on the photon to nil value. It is the inherent property of the 2D energy fields to move all disturbances in it (even in the form of 3D matter particle) at the highest possible speed. Ability of the ejection force to overcome resistance determines this highest possible speed, which we observe as the speed of light.

Linear and angular speeds of a photon are with respect to the 2D energy fields. Its linear speed is a critical constant; because that is the highest possible linear speed, it can move under the transfer of distortions in the 2D energy field, without breaking down 2D energy field's latticework structure. An attempt to increase photon's linear speed tends to increase its matter content (by assimilating quanta of matter from surrounding 2D energy fields) with associated increase in photon's spin speed rather than increasing its linear speed. An attempt to reduce photon's linear speed tends to reduce its matter content (by discarding quanta of matter into surrounding 2D energy fields) with associated reduction in photon's spin speed rather than reducing its linear speed. Hence, linear speed of light in any region space, in any direction is a critical constant. A photon traverses the same number of 2D energy fields' latticework squares in a plane in the same intervals of time. (Note that the scale of both time and distance are defined in terms of observed speed of light or similar references).

Usually, the observer is also located in the region of 2D energy fields, where the speed of light is considered. Under such conditions, linear speed of light with respect to the observer is identical in all directions, irrespective of motions of the observer. This is because the linear speed of the observer with respect to the surrounding 2D energy fields is negligible, when compared to the linear speed of light with respect to the surrounding 2D energy fields. [A fish, floating in a water-current observes any other body, moving with respect to water current, as moving at its true relative speed with respect to the current, irrespective of the direction of its motion. If this relative speed of the bodies is constant with respect to the water-current, all objects within the current and moving with respect to the current appear to move at constant speed irrespective of their directions of motion. Relative speed of the fish and other moving objects within the current will come into prominence only when the fish is able to move with a speed comparable to the speed of moving objects, with respect to the current].

If the observer is small enough and move with considerable speed with respect to the surrounding 2D energy fields, linear speed of light in the region will obey all rules of relative motion, as any other body's motion. Discrepancies appear only when the speed of light in different regions of space (with different 2D energy fields distortion status) are compared. This is how we came to regard the linear speed of light as different inside and outside a transparent medium. When the light is inside a transparent medium within the region of the observer, speed of light is considered to be lower and the scale of time is considered as constant. When the light is outside the region of the observer, speed of light is considered to be constant and the scale of time is considered to have expanded.

4.5. Distortion fields:

Gravitational actions, essentially, requires a gap in the 2D energy field. Gravitational effort/pressure is applied by the 2D energy field on to a disturbance (even if the disturbance is in the form of 3D matter-body) within the gap. Gravitational effort is applied on to a disturbance as long as it is in existence in the 2D energy field. Due to continuous application of gravitational action, 2D energy field-latticework squares surrounding a disturbance remain distorted as long as the disturbance is in existence. Directions of the 2D energy field-distortions are curved, with their concave curvature towards the centre of curvature of disturbance's perimeter.

When number of basic 3D matter particles (photons) form a union of composite matter-body, their inertial pockets combine to form a distortion field about the composite body. Distortion fields of smaller body-parts together form 'matter field' of a macro body. Depending on the nature of distortions in a distortion field, it can have different characteristic properties.

Transfer of distortions in 2D energy fields moves a matter-body. Conversely, movement of a matter particle through a 2D energy field can produce distortions in the 2D energy fields. Two sets of 2D energy field-distortions, in different directions about a matter particle tend to move the particle in a resultant direction, from higher distortion-density region to lower distortion-density region. Movement of the matter particle in the resultant direction produces distortions in the 2D energy fields in the direction of motion of the particle, while the original 2D energy field-distortions, which caused particle's resultant motion, are lost to the particle due to its displacement from the direction of transfer of the distortions. State of motion of the particle in the new direction is maintained by the 2D energy field-distortions, caused by the motion of the matter particle in the resultant direction. A matter particle, moving under more than one set of 2D energy field-distortions, cause production of independent set of distortions, corresponding to its current motion, in the surrounding 2D energy fields.

Distortion field of a matter particle is a local region in the 2D energy fields inside and outside the border of the particle. It does not require a discontinuity in the 2D energy fields. Due to the latticework structure of a 2D energy field, distortions in it, have to form a closed loop, unless the distortions are radial in nature. If the distortion starts at a point, it has to spread through the 2D energy field and return to the starting point, so that there is no discontinuity in the latticework structure of 2D energy field. Development of a distortion field is an inertial action. Unlike the 2D energy field-distortions, which act on matter-bodies due to gravitation, 2D energy field-distortions in a distortion fields cannot act on matter-bodies, because certain actions of the matter-bodies are the cause of the distortions, which develop the distortion field. A distortion field has no origin or terminus at the border of the matter-bodies. Overlapping two distortion fields change the 2D energy field distortion-densities on either side of a matter-body. Tendency of 2D energy fields, to achieve homogeneity, tends to move the distortions from higher distortion-density region to lower distortion-density region. Transfer of distortions in the 2D energy fields carries the matter particles, which are producing the overlapping distortion fields, to move them in space. This phenomenon appears as the attraction or repulsion between the matter particles. Displacement of matter-body in space is an inertial action. During this motion, additional 2D energy field-distortions are created within the matter field of the bodies to change their state (of motion).

A region of 2D energy field, occupied by work in the form of distortions, may be called a distortion field. Magnitude of distortions in unit area is the 'distortion-density'. Absolute distortion density of all 2D energy fields is the same, irrespective of their magnitude of distortions. More distortions in a part of a 2D energy field do not increase its distortion density. This is because of the reduction in the scale of distance measurements used. As the magnitude of distortions in a region increases, scale of distance measurement decreases correspondingly. Only when we compare one region of 2D energy field with another, different relative distortion-densities appear. This is because the scale on distance measurements in each region of space is unique to that region. For all practical explanations in this text, distortion density of a region of 2D energy field is expressed in terms of absolute scale of distance measurement that is suitable for undistorted part of a 2D energy field – the free space. Depending upon the nature and actions of the distortions, the distortion fields may be further classified into various types, like; gravitational, electric, magnetic, nuclear, etc. fields. All these distortion fields are static with respect to the elements, which produce them. Only, when two or more of the distortion fields overlap (interact), they can cause translatory motion of distortions/work in 2D energy fields. A moving region of work/distortion field in the 2D energy fields may be called an 'inertial field'.

In order to simplify the explanations, complicated nature of 2D energy field-distortions in a distortion field in a plane, we may resolve the natures of distortions in a 2D energy field into various components. There are three possible varieties of distortions in a 2D energy field latticework – linear, angular and radial. Imaginary lines of forces with arrows on them may be used to indicate directions of components of a distortion field. These lines of forces are also used to indicate possible direction of inertial action that may be caused by a distortion field. If linear directions of two components of interacting distortion fields are in opposite directions, they tend to neutralize each other. If linear directions of two components of interacting distortion fields are in the same direction, they tend to enhance each other. Depending on the relative linear directions of lines of forces, angular distortion fields also tend to enhance or neutralize each other. However, two radial distortion fields are unable to enhance each other. Inertial actions due to interaction between distortion fields are interpreted as attraction or repulsion between matter-bodies, which are associated with the distortion fields.

4.5.1. Linear distortion field:

In this type of distortion fields, 2D energy field latticework squares are compressed or expanded in the same linear direction. This gives rise to magnetic nature of a distortion field. 'Magnetic fields' are linearly distorted region of 2D energy fields. Since there are no bodies that naturally produce linear distortion fields, magnetic nature of a distortion field can be produced only by arranging number of angular distortion fields, in suitable array. End of a linear distortion field, from where the lines of forces appear to come out (of the body or region producing the linear distortion field), is called the North magnetic pole and the end of a linear distortion field, to which the lines of forces appear to enter is called South magnetic pole. A small part of a curved line of force, with low curvature, acts as a linear line of force. Hence, an angular distortion field, where its lines of forces have less than certain magnitude of curvature, acts as linear distortion field.

4.5.2. Angular distortion field:

2D energy field-latticework squares are distorted in angular direction. Lines of forces are curved lines with arrows in clockwise or anti-clockwise direction. This gives rise to electric nature of a distortion field. An 'electric field' is angularly distorted region of 2D energy fields. Constituent photons in primary particles (bitons) move in circular paths. Hence, all primary particles and all superior particles (they are unions of bitons) have electric fields. Due to the angular nature of electric field, its lines of force are circular lines in the (resultant) direction of motion of photons in the primary particles. Looking from one side, the lines of force appear in clockwise direction. This side of the electric field is the 'positive electric charge'. Looking from the opposite side, the lines of force appear in anti-clockwise direction. This side of the electric field is the 'negative electric charge'. Electric charges are relative directions of an electric field. Since they are relative directions, electric charge of an electric field depends on the reference used. Positive and negative electric charges have no separate or independent existence as is believed today. Every electric field has both positive and negative electric charges. They are similar to north and south poles of a magnetic field.

Field forces or inertial actions on corresponding bodies, produced by the interaction between electric fields, not only depend on the type of electric charges but also on the distance between them. At certain distance (zilch force distance, which is less than the radius of an electronic orbit in an atom) between two electric fields, they produce no field forces or inertial motions of the corresponding bodies. Beyond zilch force distance, due to lower curvature of lines of force, magnetic nature of the distortion fields dominate and the electric fields behave like magnetic fields. Electric nature of angular distortion fields (during interaction between two angular distortion fields) are exhibited only when the distance between them is less than zilch force distance, where their lines of force have greater curvatures.

4.5.3. Radial distortion field:

2D energy field latticework squares are distorted in linear directions, radially towards or away from a central point. This type of distortions gives rise to a 'nuclear field'. Only two types of fundamental particles – positrons and electrons - produce nuclear fields. If the radial distortions are directed outwards from the central point, they produce repulsive nuclear field. Fundamental particles, associated with repulsive nuclear field (the electrons), apparently repel all other primary and fundamental particles. If the distortions are directed inward, towards a central point, they produce attractive nuclear field. Fundamental particles, associated with attractive nuclear fields (the positrons), apparently attract all other primary and fundamental particles.

4.6. Macro Bodies:

All superior matter-bodies are made up of basic 3D matter particles - the photons, in various combinations. Although the processes of their development are similar, depending on relative arrangements of photons within these bodies, they exhibit diverse characteristic properties. A very brief description on the development of superior matter particles by the photons is given below.

Two (complimentary) high-matter content photons, under suitable conditions, combine to form a binary unit of spinning bodies (spinning about a common axis) moving in a circular path about a common centre. This unit is a primary particle called 'biton'. Binding force is provided by the apparent gravitational attraction between the constituent photons. Due to the curved paths of photons in bitons, their inertial pockets are permanently in unstable states. Unstable inertial pockets of constituent photons combine to form 'distortion field' of the biton. 2D energy field-distortions in the distortion field of a biton are angular in nature. Hence, all bitons have primary 'electric fields' about them. Bitons are self-sustaining matter-bodies. Bitons, in turn combine to form fundamental particles, atoms, molecules, etc., to form macro bodies. Each particle of a macro body has its constituent photons and associated inertial pockets. Distortion field of a superior particle is the resultant of distortion fields of all its constituent bitons.

Two complimentary bitons, under suitable conditions, move towards each other under apparent gravitational attraction and combine to form a union (single composite body). Photons in both the bitons move in circular paths about the common centre. This is a self-sustaining primary matter particle – 'tetron'. Bitons in a tetron are in mutually perpendicular planes. Hence, distortion fields of constituent bitons are unable to form a resultant distortion field about a tetron. Free tetrons in space, under apparent gravitational attraction, join to form layers, which folds upon themselves to create spherical tetron-shells. A completed tetron-shell has no resultant distortion field about it. This body of tetron-shell is a 'neutron'.

Three complimentary bitons, under suitable conditions, move towards each other under apparent gravitational attraction and combine to form a union (single composite body). Photons in all the three bitons move in circular paths about a common centre. This is a self-sustaining fundamental matter particle – 'hexton'. Bitons in a hexton are in mutually perpendicular planes. Distortion fields of constituent bitons combine to produce resultant distortion field about a hexton. Complicated nature of resultant 2D energy field-distortions about hextons may be resolved into linear, angular and radial distortions. Differences in the directions of resolved components of 2D energy field-distortions distinguish hextons into two types: 'electrons' and 'positrons'.

2D energy field-distortions about electrons (one type of hexton) have the following resolved components: Linear 2D energy field-distortions with lines of force projected inwards to both of its poles. This endows an electron with magnetic south poles at both its ends, with no well-defined magnetic north pole. Angular 2D energy field-distortions have lines of forces projected around the equatorial region of electron. Thus, electrons have electric fields (with both positive and negative electric charges). [Electric charge is nothing but relative direction of angular distortion field]. Radial 2D energy field-distortions, concentrated near the polar regions, have lines of forces projected outwards from their central axes. This endows the electrons with repulsive nuclear fields. Due to repulsive nature of electron's nuclear field, electrons cannot form union with any other fundamental particle.

2D energy field-distortions about positrons (one type of hexton) have the following resolved components: Linear 2D energy field-distortions with lines of force projected outwards from both of its poles. This endows a positron with magnetic north poles at both its ends, with no well defined magnetic south pole. Angular 2D energy field-distortions have lines of forces projected around its equatorial region. Thus, positrons have electric fields (with both positive and negative electric charges). Radial 2D energy field-distortions, concentrated near the polar regions, with lines of forces projected towards their central axes. This endows the positrons with attractive nuclear fields. Due to attractive nature of positron's nuclear field, positrons readily form union with other fundamental particles.

Tetrons may form a layer around the equatorial region of a positron. As and when such a layer is completed to fold upon itself into a spherical shell, it forms a neutron-like particle with a positron in its shell. This union of matter particles is a 'proton'. Protons always find electrons to form planetary systems of hydrogen atoms.

If tetrons are available in greater numbers, it is more probable for positrons to grow two tetron-layers, simultaneously, about their equatorial regions. These layers bend in opposite directions into two spherical tetron-shells about a common positron. Such a union of matter particles is a 'deuteron'. Deuterons (presently counted as one proton plus one neutron) are major components of nuclei of atoms. Deuterons, under apparent gravitational attractions and different field forces, combine to form spinning nuclei of different types of atoms. Occasionally, protons and neutrons are also included in the structure of nuclei. Nuclei of atoms are tubular in structure with varying girth at different sections. Nuclei of atoms capture available free electrons from surroundings to complete formations of atoms. Each electron is paired to a positron in the nucleus and orbit around the nucleus under the central force provided by (only) the apparent gravitational attraction.

Distortion fields about most types of atoms have resultant 2D energy field-distortions about them. Due to the field forces produced by these resultant distortion fields and apparent attraction due to gravitation, different atoms form unions until the unions are left with no resultant distortion fields. These types of unions of atoms are the molecules. Large numbers of molecules integrate into macro bodies. A macro body contains sufficient 2D energy field-distortions in and about its body-dimensions to sustain the integrity and current state of its constituent particles and the macro body itself.

4.6.1. Matter field:

Distortion fields of all matter particles in a macro body, together, form the macro body's 'matter field'. Matter field of a macro body contain enough 2D energy field-distortions in it, to sustain the stability and integrity of its matter particles and the combined body in its current state of motion. Due to the latticework structure of the 2D energy fields, matter field of a macro body extends outside its body-dimensions. Magnitude of this extension depends on the size of the macro body and the distortion-density of its matter field. Distortion-density, in the matter field, gradually reduces outwards; from the macro body's perimeter until all distortions are lost and the latticework squares of matter field becomes undistorted latticework of 2D energy field in free space. This region around a macro body, where the matter field of the body gradually loses its characteristic 2D energy field-distortions may be compared with the 'gravitational field' of the body.

In a static macro body, its matter field contains all 2D energy field-distortions required for stability and integrity of its constituent matter particles and the composite body. Additional 2D energy field distortions, introduced into the matter field of the macro body from an external source, induce macro body's whole-body motion. This part of 2D energy field-distortions is usually considered as the 'work' done on the macro body. Rate of this work with respect to change in the linear speed of the macro body is the 'force'. Matter particles of the macro body move with respect to the 2D energy fields. Although the 2D energy fields are steady in space, it is the transferred distortions in them, which are moving the matter particles of the macro body. Because of this arrangement, even though the matter particles of a macro body are moving with respect to static 2D energy fields, no resistance is offered by the universal medium of 2D energy fields to the motion of the body-particles. A macro body, moving through the 2D energy fields, does not suffer resistance/drag from the medium. In this sense, 2D energy fields behave like an ideal fluid.

Distortions in the 2D energy fields move in straight lines, separately in each of the planes. Linear motions of body-parts in different directions at different linear speeds produce rotary motion of a macro body. If the constituent matter particles of a macro body are moved away (by another force) from the linearly moving 2D energy field-distortions of a linearly moving macro body, linearly moving distortions will be lost from macro body's matter field into space and the macro body will stop responding to the lost distortions. State (of motion) of a macro body depends on the distortion-density of additional distortions (other than the distortions required to sustain the integrity and stability of the macro body and its constituent particles) and the distribution of the additional distortions in its matter field.

Introduction of 2D energy field-distortions from external source into the matter field and their stabilizations about the macro body, takes time. This time delay gives rise to the phenomenon of inertia, which is presently attributed to the body's mass. Inertia is a property of associated matter field of a matter-body (2D energy fields). Matter content of a body is inert and mass is nothing but a mathematical relation. It is the associated 2D energy fields, which produce all apparent actions/interactions, presently attributed to the masses of matter-bodies. Once, certain magnitude of distortions are introduced into the matter field of a macro body, it remains permanently within the matter field and continues keep the macro body in its current state of motion indefinitely, until the distortions are lost or removed (neutralized by distortions in opposite direction) by an external effort. Since the additional distortions (introduced by external source and moving the matter particles) in a matter field are associated with the matter particles, speed of their transfer through 2D energy fields is limited by the magnitude of additional distortions in the matter field. Hence, a macro body may move at any speed, lower than the highest permitted speed by the 2D energy field (less than the speed of light).

As the speed of a macro body approaches the speed of light, constituent matter particles of the macro body break-down to inferior particles until the macro body's speed reaches the speed of light. At the speed of light, only photons from the macro body survive. Beyond this speed no matter particle can move. This phenomenon limits the speed of 3D matter-bodies in space to less than the speed of light. Gradually, even the photons revert back to quanta of matter in the 2D energy fields. Continuous recycling of matter between 3D macro bodies (where the entropy increases) and 2D energy fields (where high order is maintained) keeps the entropy of universe within limits. Total magnitude of matter in the form of 3D macro bodies in the universe vary cyclically.

Inertia is a property of 2D energy fields, produced by their latticework structure. Apparent gravitational attraction is the product of difference in the extent of 2D energy fields on opposite sides of a 3D matter-body. Both these phenomena have nothing to do with mass of a body, which is the mathematical relation between an external force on a macro body and the body's acceleration. Hence, differentiation into gravitational mass and inertial mass is arbitrary.

4.6.2. Motion of macro bodies:

Distortions, introduced into one of the 2D energy field latticework squares, cannot remain isolated. Because of the inter-linking of the squares in the latticework, at least part of the distortion in one latticework square is transferred to the adjacent latticework squares in the same 2D energy field. Distortions, produced by an effort on any part of the 2D energy field, are progressively absorbed by adjacent latticework squares, allowing them to be strained and distorted. Latticework square, nearest to the point of action of the effort is distorted by highest magnitude, the latticework, next in front is distorted to a lesser degree, latticework square, next in front is distorted to still lesser degree and so on. This is how distortions created by an effort (force acting on a part of the 2D energy field) are transmitted through the 2D energy field. 2D energy field, as a whole, remain stationary in space, while it is the distortions (containing the work) in it, which are transmitted. A latticework square move only so much as required to store the work of its share. Rest of the work is transferred to the next latticework square and so on. Force is the rate of investment of (additional) work/distortions in 2D energy field. During transmission of distortions, each square of the 2D energy field latticework absorbs part of the work-done by remaining in distorted condition to certain degree and passes on the rest of distortion to the subsequent latticework squares. They become free to return to their stable condition only on removal of the force-applying mechanism.

Since a 2D energy field extends only in one plane, no distortions can be transmitted directly into the third spatial dimension of the 3D space system. Transmission of distortions is restricted to the plane of the corresponding 2D energy field. A 3D matter particle simultaneously occupies gaps in many 2D energy fields (3D space) in the same location. 2D energy field-distortions, acting on a 3D matter particle in a plane, move it. The particle being three-dimensional, during its motion it produces distortions in all other 2D energy fields occupied by it. In this way, distortions in one 2D energy field may be indirectly transferred or transmitted to other 2D energy fields. An effort, (presumably) acting, through the medium of 2D energy fields, on a 3D matter particle has its components in one or more of the 2D energy fields, in the planes occupied by the particle. Each 2D energy field transmits distortions only in its plane. Such actions by various 2D energy fields (occupied by the 3D particle), together, produce a straight-line transmission of distortions in 3D space system. Thus, it appears that a force/work in 3D space system is always transmitted in a straight line.

In figure 4F, ABDC (shown in dotted lines) is an undistorted 2D energy field-latticework square. All junction points of the undistorted latticework square are under equal external forces from the latticework. Consider a hypothetical external effort, 'F', as shown by the arrow, acting on the latticework square at the junction point 'C'. Junction 'C' is displaced to 'C_2'. Other junction points, though not under external efforts are free to move. They are displaced as per physical laws applicable to latticework structures. Distorted position of the latticework square is shown by $A_1B_1D_1C_2$. Distortion in the latticework square continues to increase at a rate corresponding to the external effort. Due to latticework structure of the 2D energy field, distortion of the square is transmitted in the direction of the external force. Latticework square nearest to the point of action of the effort has highest magnitude of distortion.

Now, let the external effort cease. Since the external effort is not present, any more, reactions at the junction points tends to reduce distortions and bring the latticework square, back to its stable state. Under reactions, junction points 'A_1' and 'D_1' tend to move inwards and junction points 'B_1' and 'C_2' tend to move outwards. Junction point 'C_2' moves at a faster rate than the junction point 'B_1'. Let $A_1B_1D_1C_1$ be the resultant shape of the latticework square after the removal of external effort. The latticework square has not regained its stable state but it remains permanently distorted to certain extent. Magnitude of this permanent distortion is the work-done by the effort and its rate is the magnitude of force during its action. Due to latticework structure of the 2D energy field, a distortion cannot remain static in space; it will be transferred in the direction of the external effort. As this permanent distortion is transferred through the latticework, matter particles, associated with them are carried with the distortions. When all matter particles

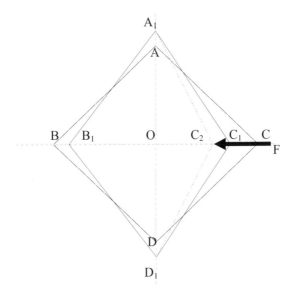

Figure 4.F

of a body are moved by distortions in their respective 2D energy fields, this creates the motion or displacement of the macro body, whose constituents these matter particles are.

After removal of the external effort, distortions in the quanta of matter A_1C_2 and D_1C_2 are reduced to bring them to positions A_1C_1 and D_1C_1, respectively. Differences between these positions (distance between junction points C_2 and C_1) is the accelerating component of work, available during the action of effort. 2D energy field-latticework structure reaches a stable state (with permanent work invested in it) only after the accelerating component of distortion ceases to exist. Thereafter, only stable permanent distortions will be left in the 2D energy field. These permanent distortions in the latticework will continue to be transferred at a steady speed. Total magnitude of distortions in the latticework remains steady until they are removed (partially or fully) by distortions introduced by another external effort, acting in opposite direction. This is the mechanism of inertial motion of a body.

Figure 4G, shows four of 2D energy field latticework squares, A, B, C and D in a straight line. G, H, K, M and N are the junction points associated with these latticework squares. An external effort 'F' acts on these latticework squares from the right, as shown by the arrow. The latticework squares are distorted from the right to left as shown. Reaction developed at any junction point is proportional to the distortion of quanta of matter at the junction point. During the action of the effort, as the latticework squares are distorted, latticework square 'A' experiences greater reaction at the junction point on the right 'N' and lesser reaction at the junction point on the left 'M'.

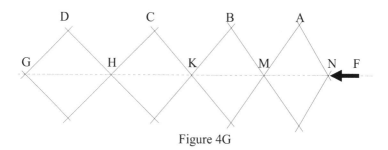

Figure 4G

Latticework square 'A' is distorted towards the junction point, M, with lesser reaction. It is distorted to the left. Arms of quanta of matter to the left are turned by lesser angle compared to quanta of matter, forming the arms to the right. Other, latticework squares are also distorted similarly to the left. Magnitude of distortions and the number of latticework squares distorted in straight-line (extent of distortion field) increase as the external effort continues to act.

When the effort ceases, further investment of distortions into the 2D energy field-latticework also comes to a stop. Part of the distortions, already introduced into the latticework structure, remains permanent in the 2D energy fields and continues to be transferred in the same direction, creating a moving distortion field. Since, there is no effort applied from the rear, distorted latticework squares are now free to expand rearward also. Part of the distortions in the latticework structure is nullified by rightward expansion of the latticework squares. This nullified part of the distortions is the accelerating component of the external effort, whose magnitude becomes zero after the inertial delay (period).

Figure 4H, shows the absolute condition of the latticework squares after removal of the external effort. 'G, H, K, M and N' are the junction points associated with latticework squares 'D, C, B and A'. Outer ends of latticework squares 'D' and 'A', junctions 'G' and 'N', respectively, have no distortions and they are part of undistorted 2D energy fields. Since the external effort is removed, reaction at junction 'N' becomes nil. Junction point 'M', being within the distorted part of the 2D energy field, has certain reaction

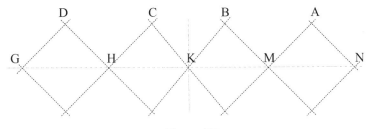

Figure 4H

in it. Latticework square 'A' expands in the direction of lesser reaction – to the right.

Similarly, all distorted latticework squares of the 2D energy field, readjust their distortions to reach a stable state.

Distortions in the latticework structure tend to be spread outwards from junction point 'K'. Let the junction point 'K', in the middle of the distorted latticework squares, has highest magnitude of distortions after the readjustments. Quanta of latticework squares 'C and B', facing the centre line through 'K', have highest magnitude of displacements of all the quanta of matter in the distorted region of 2D energy field.

Junction point 'H' has lesser distortion than junction 'K'. Latticework square 'C' strains to expand. In order to expand, it has to apply greater reaction to the right and it needs to apply lesser reaction to the left. Latticework square expands in the direction (to the left) towards, where it requires less reaction. Square 'C' expands to the left. Junction 'H' is displaced to the left, with respect to junction 'K'. Distortions of quanta of matter at the junction, which are part of the latticework square 'D' and hence the distortions in the square 'D' increases. Part of this distortion is now transferred to the latticework square to the left of 'D'. Similarly, distortions in the latticework squares, to the left of junction 'K' are transferred to the left. Undistorted latticework squares to the left of the region are distorted. This process transfers the work invested in the region, in leftward direction within the latticework structure of the 2D energy field.

Expansion of latticework square 'C', to the left, reduces strain in its arms at junction point 'K'. Latticework square 'B', which tries to expand to the right, now, has lower strain at its leftward junction point 'K'. Anchor point of square 'B', to expand itself, is now changed from junction point 'K' on the left to the junction point 'M' on the right. Latticework square 'B' expands to the left instead of to the right, as shown figure 4H. Although the square 'B' expands as shown in figure, junction point 'M' does not shift from its location, as shown in figure 4G. Expansion of the latticework square is realized by the shifting of junction point 'K' to the left. Similarly, the latticework square 'A' also expands to the left and keeps the junction point 'N' in place.

Distortions in the latticework squares, with respect to 2D energy fields, outside the region of transfer of work, are as shown in figure 4H. Highest magnitude of distortion density is at the rear region of distorted latticework squares. Towards the forward end of the region, distortion density reaches minimum magnitude, the latticework squares just outside the region has no distortions and they are parts of undistorted 2D energy field. Similarly, latticework squares outside the region of transfer of work to the rear has no distortions in them.

Distortions in the latticework squares, with respect to each other, remain as shown in figure 4G. They are in their steady states after removal of accelerating component. These distortions in the 2D energy field are transferred in forward direction without changes in relative distortion-densities. With respect to the latticework squares, distortions in the latticework structure are maintained, while they are being transferred in the 2D energy field, to the latticework squares in front. 2D energy field or its latticework squares do not move along with the distortion field but only the distortions (work) in the 2D energy field is transferred.

With respect to the 2D energy field, outside the region, latticework squares in front are newly distorted and latticework squares at the rear are relieved of all distortions. They appear as shown in figure 4H. All steady state distortions are confined to the limit of the matter field of the macro body, shown by latticework squares 'C and D'. Distortions are also extended to the rear of this region, as in latticework squares 'A and B'. Distortions in latticework squares 'A and B' reduce as the matter field is transferred to the front. Matter field is constantly transferred in a straight line through the 2D energy field, which is stationery but for small movements of constituent quanta of matter in place. This is the method of transfer of a matter field (work/distortions about a macro body) through 2D energy fields.

As the distortions in the 2D energy fields about the macro body (macro body's matter field) move forward, constituent photons of the macro body are carried along with the moving distortions. Photons are moved in the direction of external effort on the macro body, without affecting their independent motions within the constituent primary/fundamental matter particles at their critical linear and spin speeds. Movement of all constituent photons in a macro body, in the direction of external effort, moves the whole macro body, as can be observed by us. Because of the composite structure of macro bodies and displacements between their constituent 3D matter particles, there are no rigid bodies in nature.

4.6.3. Inertia of rotary motion:

Forces or work may be transmitted, through the 2D energy fields, only in straight lines. Hence, all inertial motions take place in straight paths. Simultaneous motions of a body-particles, produced by different magnitudes of works in body's matter field, in same direction, may appear as a linear motion of the body in a curved path. Simultaneous motions of a body-particles, produced by different magnitudes of works in body's matter field, in different directions, may appear as a rotary movement of the body. Additional work, introduced into the matter field of a macro body, to produce its linear motion, remains within the body as long as the body's path is limited along the line of transfer of its matter field-distortions, producing the motion. Similarly, additional works introduced into the matter field of a rotating body, in various linear directions remain within the rotating body.

Consider a small external effort, applied at the center of a macro body of infinite length, in the direction of its length. This external effort introduces additional distortions in the body's matter field that tend to move the body as a whole (initially, only a part of the body) at a constant speed in the direction of its length. Additional distortions, set up in body's matter field, spread throughout the length of the body with a gradient in the distortion-density, decreasing towards the forward end of the body. As the body is of infinite length, this action of spreading-out of the additional distortions within the body, along its length will continue indefinitely, while the body (part of the body, where the additional distortions are present) continues to move. Since the body is of infinite length, forward end of the body will move only after infinite time, when the distortions in the matter field reach the forward end. In the mean time, other parts of the body are moving in the direction of the external effort.

Though this is a hypothetical condition, it is the working principle of rotary motion of a body. During the rotary motion of a macro body, about a center point, any point within the space occupied by the macro body is always within the limits of body-dimensions. Therefore, as far as this point (in 2D energy field) is concerned, the macro body is of infinite length in the direction of motion of the point. Additional distortions of the matter field, spreading-out towards the forward direction, never reach the end point of the body; because in a circular body there are no end points. Therefore, a spinning body will continue to spin at constant angular speed indefinitely, until another external effort acts on it.

Uneven action of a linear effort about the centre point in a macro body produces a 'couple of force' (torque) and its resulting rotary motion (along with any linear motion produced by the effort). Rotating motion of a body is nothing but sum of linear motions of its 3D matter particles, moving at different linear speeds and in different directions about a centre point. These motions are over and above natural motions of these particles, required for their

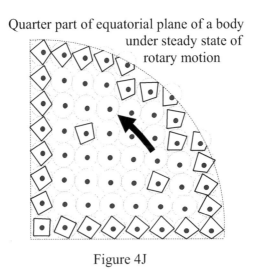

Quarter part of equatorial plane of a body under steady state of rotary motion

Figure 4J

sustenance. With respect to a radial line of the rotating macro body, (tangential) linear speeds of molecules increase in proportion to their distance from the centre of rotation. Unequal linear speeds of transfer of resultant distortions move matter particles, in the region, corresponding to the spin speed of the body. With respect to rotary motion of a body (that has no linear motion) resultant matter field distortions, producing body's rotary motion, remain steady in space and the body itself acts as if it is a linearly moving body of unlimited length with respect to each point in the body. Matter particle at the centre of rotation has no resultant matter field-distortions about it and hence it does not move during rotation of the macro body. Matter particles, whose distances from the centre of rotation are in opposite directions, move in opposite linear directions.

Figure 4J shows a quarter part of a plane (perpendicular to the spin axis) of a rotating body. Black circles represent composite 3D matter particles (molecules) and the circles around them in dotted line show the limits of their distortion fields. Distortions in the matter field of the macro body, required to maintain the state of linear motion and integrity of the body, are ignored. Rectangles in thick lines around few of the matter particles show representations of deformations of latticework squares in the resultant distortion fields of the molecules. The thick curved arrow shows direction of rotation of the body.

4.6.4. Motion due to torque:

In order to produce a torque, two parallel (non-concurrent) efforts (or their components) acting in opposite directions are necessary. In case of a supported body, one of the efforts may be supplied by external means while the other effort is provided by the reaction from the supporting pivot. In very large free macro bodies or if the external effort is of high power, centre of mass of a macro body may act as body's support.

To analyze the rotary motion of a macro body, produced by a torque (an external effort applied about a point at a distance away from the point of support and in a direction other than along the line passing through the point of support), we shall consider a thin ring supported and pivoted at its geometrical center. Let a linear effort act on the ring, at a point on its rim, in tangential direction. 3D matter particles of the body, at this point, tend to move in the direction of the external effort. Since these particles are bonded to other matter particles in the ring, interactions (field forces) develop between them due to the intended displacement between them. These interactions act on each of the matter particles in the body to move them in appropriate directions, so that the integrity of the body is maintained.

Interactions on each of the matter particles act as external efforts on them, to induce their inertial motion. Thus, additional 2D energy field-distortions are developed continuously throughout the matter field, within the ring, in appropriate directions. Only those distortions, introduced at the point of application of the external effort are directly induced by the external effort. We will consider matter field-distortions induced at the point of application by the external effort for further explanation. Additional distortions, introduced by the movements of body's matter particles, in the matter field of the body in various directions also operate in similar manner.

Inertia is the property of 2D energy fields (matter field of a macro body). Distortions in the matter fields are able to travel only in a linear direction and in the same plane. Taking a small part of the rim, we can see that, as the macro body turns, the point (3D matter particle) influenced by the additional distortions moves away and out of influence of these distortions, which are being transferred in straight line. Matter field-distortions move in straight line but the 3D matter particle is carried along with the body of the ring in a circular path, as the ring turns. Therefore, the matter particles at a point on the ring are bound to go out of the line of action by the additional distortions, introduced by the external effort. Before these additional distortions can escape from the body-limit, other matter particles in the body are brought into the place of the original particles and the additional distortions in the matter field are now interacting with newly placed 3D matter particles. In other words, due to the rotary motion of the ring, additional distortions in the matter field are supplied with new matter particles continuously, to interact with. Magnitude of the additional distortions at any point in the body of ring is sustained by each matter particle, producing appropriate additional distortions to act on the matter particle, moving in front of it. Although the additional distortions, introduced by the external effort, are long lost from the matter field of the rotating macro body, movements of all its matter particles, together, create and sustain the total additional distortions in the matter field of the macro body.

The ring acts as a body of infinite length at the point of application of the external effort. The ring continues to accelerate (angularly), as long as the external effort is acting on it. All the additional distortions, introduced by the external torque are stored within the matter field of the ring as linear additional distortions in various directions, required for its rotary motion. They are superimposed on any other distortions, already existing in its matter field. After the external effort is withdrawn, additional distortions in the matter field, all around the ring, stabilize and the ring continues to rotate at constant angular speed. Additional work-done in the body is sustained constant.

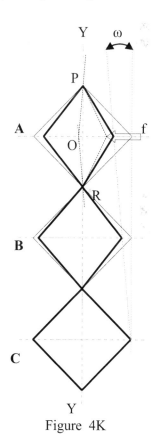

Figure 4K

Action of a rotating force:

Figure 4K shows representations of three matter field latticework squares A, B and C of a (rotating) macro body, in a radial line. Dotted lines show their original shape, when the macro body has no motion at all. [Matter field-distortions, maintaining the steady state of rest and integrity of the body are ignored]. Latticework square A is near the perimeter, C is at the centre of the rotation and B is somewhere in between the centre of rotation and the perimeter of the macro body. In their original shapes, arms of all latticework squares are symmetrical about a reference line YY (vertical centre line, shown passing through the centers of the squares). Due to this symmetry, an external effort acting along the reference (radial) line is evenly distributed in the arms of the latticework squares and the resultant action of the effort is linear along the reference line.

Another external linear effort, f, acting on the macro body away from the centre of mass of the macro body and perpendicular to the reference line, distorts the matter field squares in the region of its action. Although the external effort distorts only the matter field latticework square A, directly, all other latticework squares are also distorted due to their mutual linkage. Magnitudes of distortions in the latticework squares, A, B and C are not equal. Differences in the distortions of latticework squares, produce spin motion of the body. Matter field squares are distorted as shown by the thick black lines, during the stable state of constant angular speed. Dashed lines shown in the square A shows the additional distortion during the acceleration period. Elongations of the latticework squares, in perpendicular direction, due to the distortions are not shown in the figure. Latticework square C, being at the centre of rotation of the body, is not affected by the external effort, f. Hence, it is not distorted. Due to integrity of the body, square B is distorted to a magnitude in between the distortions of latticework squares A and C, corresponding to its location.

As the location of matter field latticework squares approach the perimeter of the body, magnitude of distortions in them approach the highest value. Speeds of transfers of 2D energy field-distortions are proportional to the magnitudes of distortions in each latticework square. Outer most matter field latticework square transfers the distortions fastest and speeds of transfer of distortions in other latticework squares gradually diminish as their location approach the centre of rotation. Matter particles near the perimeter of the rotating body have highest linear (tangential) speed and linear speeds of matter particles nearer to the centre of rotation are lower. Due to different linear speeds of the particles, the body as a whole rotates about its centre of rotation. Part of the body between the centre of rotation and the point A on the periphery moves in the direction of external effort and the part of the body on the opposite side of centre of rotation moves in the opposite direction. Motion of one part of the body, in opposite direction to the external effort, is produced by the field forces of integrity in the macro body.

Unlike in the cases of linear motion, distortions in the matter field latticework squares of a rotating body, about the reference line, are different for each matter field square. Magnitude of this difference is the angular speed of the rotating body, shown in figure 4H as angle ω. When the external effort is withdrawn, the macro body will attain steady state of rotation, after the inertial delay. States of distortions in the matter field latticework squares, during the steady state of rotary motion, are as shown by the thick lined squares in the figure. Macro body will maintain this state of rotation at constant spin speed until another torque is applied to modify its angular speed.

Differences in the magnitudes of distortions, between matter field latticework squares along a radial line, create asymmetry in their arms with respect to the radial line YY. Due to the asymmetry of the arms of matter field latticework squares, action of an external effort along the radial line is bifurcated into unequal components. One component of action produces an angular deflection in the direction of rotation to turn the macro body and the other component imparts linear motion to the body along the radial line. Direction of resultant action of the external effort is deflected from the direction of its application by an angle whose magnitude is proportional to the body's angular speed of rotation.

In order to produce pure linear motion of a rotating body in any desired direction, the external effort has to act along a line deflected away from the geometrical radial line of the rotating body, in the desired direction. This is the radial line of the matter field, corresponding to the symmetrical states of arms of its latticework squares in the matter field of the rotating body. Hence, there is an angular difference between the direction of action of an external effort and the direction along which it is applied on a rotating body. This phenomenon shifts the direction of tides on a spinning body from local meridians, where the external effort appears to act.

Action of linear force on rotating body:

External effort, acting through the center of a rotating macro body (in motion or in steady state) in the plane of its rotation, introduces its own additional distortions in the matter field of the body. These distortions are in the direction of the effort (force) and they modify the additional distortions in every part of the body. Matter particles in any part of the body will have a corresponding linear motion, in addition to its rotary motion. Resultant distortions, acting on the matter particles, are modified to endow them with motion in the resultant directions. Changes in the matter field of the body cause shift in the instantaneous center of rotation of the macro body. During this unstable period, rotation of the macro body will become eccentric and if the body is free, it tends to rotate about its instantaneous center of rotation, which is away from its centers of gravitation and mass. At any instant, matter particles of the body will be under acceleration and deceleration according to their relative position within the body.

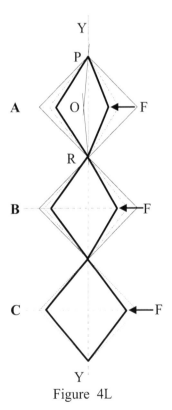

Figure 4L

Figure 4L shows representations of three matter field latticework squares A, B and C of a (rotating) macro body, along a radial line. Dotted lines show their original shape, when the macro body has no motion at all. [Matter field-distortions, maintaining the steady state of rest and integrity of the macro body are ignored]. Latticework square A is near the perimeter, C is at the centre of the rotation and B is somewhere in between the centre of rotation and the perimeter of the macro body. In their original shapes, arms of all latticework squares are symmetrical about a reference line YY (vertical centre line, shown passing through the centers of the squares). Due to this symmetry, an external effort acting along the reference (radial) line is evenly distributed in the arms of the latticework squares and the resultant action of the effort is linear along the reference line.

Action of a torque rotate the macro body about its centre of rotation, represented by matter field latticework square C. All other matter field latticework squares of the body will be correspondingly distorted, as shown by the dashed lines in the figure.

Let an external effort F, as shown in figure 4L, act evenly on the macro body, which is rotating at a steady angular speed. And the body attains a steady state of linear motion in addition to its spin motion. All matter field squares in the body are distorted by identical magnitudes due to the external linear effort. Squares in dotted line show the original state of latticework squares in static state of the body (inherent distortions are ignored). Squares in dashed lines show the distortions, causing steady rotary motion of the body. Additional linear motion of the body, produced by external effort F, has its own linear distortions in the matter field. Resultant shapes of three latticework squares A, B and C are shown in thick lines. Since whole of the matter field is linearly distorted identically, the body maintains the linear motion without affecting angular speed of its rotary motion. Asymmetry of arms of the matter field squares, angular difference between radial line, YY, and median lines of distortions, PO and RO, in the square A are enhanced in proportion to body's linear speed. Correspondingly, angular deflection of matter field distortions from the reference line is enhanced in proportion to body's linear speed. Action of another external linear force on a spinning body depends not only on its magnitude but also on the symmetry of matter field squares to the direction of external effort. This phenomenon enhances the shifts in the direction of tides from local meridian of a planet, corresponding to its linear speed.

With respect to a composite macro body, its linear and rotating motions are distinctly separate. Only an external linear effort applied evenly on the body can modify linear motion of the body. Only another torque (linear effort applied unevenly on the body) can modify a body's rotary motion. Although an external effort may simultaneously invoke linear and rotary motion of a body, additional works invested in body's matter field are distinct for each of these motions. For linear motion, work is of linear nature and for spin motion; work is also of linear nature but varying in magnitude and direction about the centre of rotation. In a macro body's steady state of motion, each nature of distortions produces respective motions independently. Even at very high linear speed of a spinning body, work corresponding to its spin motion remains latent within body's matter field and rotates the body about its centre of rotation. Transition period between one steady state of motion to another is the body's acceleration stage. During acceleration stage, external linear effort (force or torque) modifies the matter field-distortions in the body. Reshaping of matter field latticework squares during this period, to modify any one type of motion, takes place without interfering with the other type of motion of the body. Change in linear speed does not affect the spin speed and a change in spin speed does not affect the linear speed of the body.

* ** *** ** *

Chapter FIVE
GRAVITATION IN 3D SPACE

Presently, gravitation is understood by its only known manifestation of apparent attraction between matter-bodies. From the above explanations, on this concept, it may be clear that the apparent gravitational attraction is only a by-product of gravitation. Gravitation has its functions both on static and dynamical systems. Main function of gravitation is the creation and sustenance of higher-dimensional matter-bodies in nature. While doing so, simultaneous gravitational (independent) actions on different bodies also cause, what appears to be attraction between the matter-bodies.

At present, source of such attraction is assigned to a mysterious property of matter-bodies. According to Newton's theory on gravitation, it is defined simply as *'the universal force of attraction between all matter'*. On the other hand, Field theories claim that *'the acceleration due to gravity is a purely geometric consequence of the properties of space-time in the neighborhood of attracting masses'* rather than an attraction between matter-bodies. None of these theories explains gravitation's motivation or mechanism of action. Explanations in previous chapters show that the phenomenon of gravitation needs not be a mystery. Real matter particles (quanta of matter) and their natural arrangements can produce the phenomenon of gravitation. Unfortunate part is that these real matter particles and their natural arrangements are beyond observational capability of rational 3D beings. However, they may be theorized as functional entities and their logical actions can be substituted for numerous assumptions, currently used.

Although, the (apparent) gravitational attraction is only an apparent phenomenon, this part of gravitational actions is dealt within contemporary science as the real and the only one gravitational action. In this chapter, we shall explore how 'various actions and limitations of apparent attraction due to gravitation between 3D matter-bodies' are explained logically by this concept. Usually, we recognize gravitational action by its inertial actions on participating bodies. Inertial actions due to gravitation on the larger of the participating bodies are usually ignored and the total resultant inertial action is credited to the smaller of the participating bodies. Hence, we always consider that a smaller body is attracted towards a larger body due to gravity. Dealing with single action on a body, rather than with simultaneous actions on two participating bodies, helps to make mathematical treatments simpler.

Action of gravitational effort/pressure on each matter particle is independent of all other matter-bodies. Development of 2D energy field-distortions about a matter particle, which produce gravitational actions on it, is an inertial action. This takes place during the creation and development of basic 3D matter particle (photon). Thereafter, the apparent interactions between basic 3D matter particles in macro matter-bodies, due to gravitation, are instantaneous. Hence, changes in the magnitude of apparent gravitational attraction take place instantly on change of parameters or constitution of participating 3D matter-bodies. Changes in the parameters or constitution of a macro body are carried out by developing appropriate 2D energy field-distortions about that body. Gravitational actions on constituent body-particles of macro bodies change simultaneously during this development. No transfer of imaginary particles/energy from one body to another is required to produce changes in the apparent gravitational attraction between two matter-bodies. However, the inertial motions of these bodies, under the apparent gravitational attraction, are again subject to inertial delay.

5.1. Gravitation in 3D space:

In this section, a 'point' is considered as a 3D body of negligible dimensions. This point-body simultaneously exists in all the planes passing through it. It is under gravitational pressure from the 2D energy fields in all these planes. Each 2D energy field applies gravitational pressure, on the point-body, in its own plane. All the 2D energy fields, in the planes passing through the point-body, together, apply the gravitational pressure on the point-body, in 3D space system.

Consider a point-body, 'A', in free space. Free space is that part of space, where there are no disturbances/distortions in 2D energy fields, other than the disturbances/distortions considered. Extent of 2D energy field in any direction from a point-body 'A' (in all planes passing through it) is infinite. Gravitational pressure applied on the point-body by each of the 2D energy fields, in planes passing through it and extending infinitely into the space, is of highest value and this value is a constant. Similar gravitational pressures are applied by all the 2D energy fields in planes passing through the point-body. Together, they constitute gravitational pressure on the point-body in the 3D space system. Since the gravitational pressure applied in each plane is of a constant magnitude, total gravitational pressure on the point-body in 3D space system is also of (highest) constant magnitude. Let the magnitude of this constant value of gravitational pressure be 'G_2'.

In figure 5A, let 'OAB' be a spherical shell of negligible thickness, with the 3D point-body 'A' as its center, distance $AB = d$ as its radius and 'B' be a point of unit area on the surface of the spherical shell. Gravitational action on the point-body 'A' through area 'B' in the direction BA (by all 2D energy fields containing both the point-body 'A' and unit area 'B') is a fraction of the total gravitational pressure applied at the point-body 'A', through the total surface area of spherical shell OAB.

Surface area of the sphere $OAB = 4\pi d^2$

Gravitational effort on point 'A' through unit area of 'B';

$$GF = \frac{G_2}{4\pi \times d^2} \qquad (5/1)$$

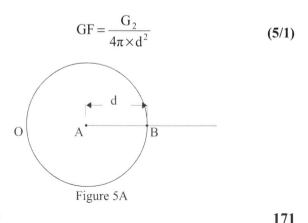

Figure 5A

It can be seen that the gravitational effort applied on point 'A' through point 'B' in 3D space is inversely proportional to the square of distance, 'd', between them.

5.1.1. Apparent gravitational attraction in 3D space:

In figure 5B, 'A, B and C' are three points in a straight line in 3D space. Let 'A and B' be two 3D disturbances (bodies) of unit measure, each. Let 'C' be a point of unit area (in a plane perpendicular to AC) and distance AB = BC = d.

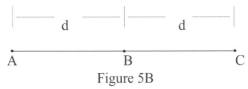

Figure 5B

Let the gravitational action on 'B' from the right, along the direction CB by each of the 2D energy fields, in the planes passing through both points 'B' and 'C' is a constant value G_1. There are more than one 2D energy fields passing through both the point-disturbances. Hence, magnitude of the gravitational action on the point-disturbance 'B', from the right along CB, by all the 2D energy fields in the planes passing through both 'B' and 'C' is a multiple of G_1 by the number of 2D energy fields in the planes passing through both of them. Let the magnitude of this gravitational action be a maximum constant of value G_2.

If the point-disturbance 'A' was not present, similar gravitational action would be applied on 'B', from the left along AB also, to neutralize the above gravitational action from the right. However, due to the presence of point body A, the extent of 2D energy fields, applying the gravitational action from the left, along AB is limited up to point 'A'. So, the gravitational action on the point-disturbance 'B' from the left along AB is equal and opposite to the part of the gravitational action from the right along CB by the 2D energy fields of their extents from points 'B' to 'C', along CB. These actions being equal in magnitudes and in opposite directions neutralize each other. They are unable to act on point 'B' to cause an inertial action. Remaining part of gravitational action on point 'B' is the gravitational effort available at point 'C', from the right towards point-disturbance 'B' along CB.

Resultant gravitational effort on point-disturbance 'B' towards the point-disturbance 'A', $\mathrm{GF} = \dfrac{G_2}{4\pi \times d^2}$ by (5/1)

5.1.2. Push gravitation:

Resultant gravitational action is on the point-disturbance 'B' and it tends to push 'B' towards the point-disturbance 'A'. That is, the resultant gravitational action on a disturbance (body) is such as to move the body away from the major extent of 2D energy fields, producing the gravitational actions. Therefore, resultant gravitational action (that cause the apparent gravitational attraction) on any of the disturbances present in the 2D energy fields is 'repulsive' in nature. Resultant gravitational action has a 'push' action on a disturbance. When there are more than one 3D disturbance in the space, resultant inertial actions of the gravitational efforts on each of these disturbances is to push them towards the other disturbance. Thus, the part of gravitational actions that produce inertial actions of bodies appear (as is wrongly assumed) to be forces of attraction emanating from these bodies.

Although the gravitational efforts (producing inertial action) on each of the disturbances is of push nature, result of their actions is to move the disturbances towards each other. External agency, producing these push actions on the disturbances is not recognized. When inertial actions on the disturbances are considered, without considering their cause, each body appears as being pulled towards the other. Even for this action, no intermediate agency is available. Under this confused state, it is simply assumed that the disturbances (matter-bodies) attract each other, by some mysterious mechanism. Since this action is generally believed to be of attractive nature (contrary to the true nature of gravitation), the same nature is used in this text with a qualification to denote the apparent interaction between two bodies, due to gravitation. Hence, the resultant inertial action due to gravitation on these disturbances is called 'apparent gravitational attraction'. Apparent gravitational attraction is a by-product of separate gravitational actions on each of the disturbances/bodies. Apparent interaction is developed because of existence of both the disturbances in the same 2D energy fields. If they do not occupy the same 2D energy fields, there will be no apparent attraction between them due to gravitation, but the actions of gravitation continue to act on both the disturbances unhindered. Even in the cases of coplanar bodies, there are certain conditions for the apparent gravitational attraction to develop between them.

With reference to figure 5B, similar gravitational effort is also available on point-disturbance 'A' towards point 'B'. Other than, for the effect of angle subtended on each other by the participating bodies, these actions are independent of each other. Total apparent gravitational attraction between disturbances 'A' and 'B' is the sum of the two (push) gravitational actions, acting separately on each of the disturbances. Addition of the resultant actions is not taken into consideration separately. It is taken care of, while determining the gravitational constant of proportion to determine the total apparent gravitational attraction between two bodies. Another reason is that, at present, the apparent gravitational attraction between two bodies is not taken as a resultant of two separate gravitational actions on the participating bodies, but as a single force of attraction between the two matter bodies, emanating (mysteriously) from both the bodies simultaneously. In the above example, both the point-disturbances 'A' and 'B' are of highest 3D matter density and all 2D energy fields in the planes passing through them are blocked by their matter content (discontinued within the space occupied by the point-disturbances).

Thus, the total apparent gravitational attraction between two point-3D disturbances;

$$2GF = 2\left(\frac{1}{4\pi} \times \frac{G_2}{d^2}\right) \qquad (5/2)$$

If the magnitude of the disturbances (considered in the example) is greater than a point, number of 2D energy fields passing through the disturbances is proportional to their matter contents. Let the matter content of disturbance 'A' is equal to m_1 and the matter content of disturbance 'B' is equal to m_2. 'd' is the distance between the surfaces of the disturbances (not the centers of disturbances).

Resultant gravitational effort on disturbance A;

$$GF = \frac{G_2}{4\pi \times d^2} \times m_1 \qquad \text{by (5/1)}$$

Resultant gravitational effort on disturbance B;

$$GF = \frac{G_2}{4\pi \times d^2} \times m_2 \qquad \text{by (5/1)}$$

Total apparent gravitational attraction between disturbances 'A' and 'B':

$$2GF = \left(\frac{1}{4\pi} \times \frac{G_2}{d^2} \times m_1\right) + \left(\frac{1}{4\pi} \times \frac{G_2}{d^2} \times m_2\right)$$

$$= \left(\frac{1}{4\pi} \times \frac{G_2}{d^2}\right)(m_1 + m_2)$$

Greater share of the total apparent attraction is the contribution of the larger disturbance.

If both the disturbances are of equal matter content, 'm' each;

$$2GF = 2m\left(\frac{1}{4\pi} \times \frac{G_2}{d^2}\right) \qquad (5/3)$$

In the above example, single basic 3D matter particles are considered as matter-bodies (disturbances). Sizes, shapes and phase relations of these particles are not considered. Instead, they are assumed as spherical matter-bodies. Larger matter-bodies are combinations of more than one basic 3D matter particles (photons).

5.1.3. Gravitational attraction between photons:

Each photon is a disturbance with respect to 2D energy fields of its existence. Gravitation can act only on the curved surfaces of a 3D disturbance. Photons have almost flat surfaces on their disc-faces. Although gravitation is applied continuously on the flat surfaces, due to constant and highest matter density of the photon, it cannot produce inertial action on them. (Refer section 3.3.3). Therefore, the gravitational actions are active only on a photon's circular perimeter. That means only those 2D energy fields passing through the disc planes of the photons contribute towards the apparent gravitational attraction between two photons. Photons with their disc-planes in different spatial planes do not develop apparent gravitational attraction between them.

A stable photon, simultaneously, spins about its axis and moves in linear direction, perpendicular to its axis. Relation between the median planes of two photons (other than in special cases) changes continuously. Apparent gravitational attraction between two photons depends on their relative orientation. In order to generate apparent gravitational attraction between two photons, it is necessary that their median disc planes (or any planes within photon's disc and containing its axis) coincide in the same plane.

If median planes of two photons happen to be in the same plane, there is an apparent gravitational attraction between them. In normal cases, the photons are extremely thin and hence their existence can be considered as limited only in their median planes. Because of its 3D existence, depending on the convex curvature of disc faces and thickness of the discs, a photon has very small gravitational action on it in other planes (which are at an angle to its disc-plane) also. Since their faces are almost flat, these gravitational actions are negligible and they are mainly used to spin the photon's matter core. As long as the internal pressure of a photon's matter-body is maintained at the stable level, gravitational actions are unable to act on its flat surfaces to produce inertial action of the photon. In the process of stabilization, small curvatures develop on the disc faces of the photons. Even for a photon of frequency of 10^{20} Hz (photons of matter contents corresponding to frequency above 10^{20} Hz are extremely rare) angle subtended by its disc-segments at the spin axis is much less than one degree of arc. Therefore, the curvature on the faces of a photon is too small to be considered, here.

Since the frequency of a photon is proportional to its matter content, time duration for all photons to spin through any one 2D energy field plane is the same. Disc sizes (diameters) of all photons, being of critical radial size, are also the same. Therefore, apparent gravitational attraction between two photons in any one plane, where their median planes coincide, is of constant magnitude (provided distance between them does not change) irrespective of their matter contents. As a result, other than in some special cases, the matter content of a photon is not a factor directly affecting the magnitude of (instantaneous) apparent gravitational attraction between two photons. However, frequency of a photon depends on its matter content. Frequency of a photon is a factor determining the number of times a photon spins through the same plane. Number of appearances in the same plane determines the average magnitude of apparent gravitational attraction, over a period. Therefore, the frequency of a photon and hence its matter content has an effect on the average magnitude of apparent gravitational attraction felt by it, over a period.

A photon (its disc plane) appears in the same plane of another photon of the same frequency and phase, once every half spin. This applies in all cases of the photons' relative orientations other than when their axes are end-to-end in a straight line. Every time two photons of the same frequency and phase come in the same plane, they apparently attract each other due to gravitation. Average magnitude of apparent gravitational attraction between these photons depends on their identical spin speeds. That is, the average magnitude of apparent attraction is proportional to photons' frequency, which in turn depends on their matter contents – rest mass. This average apparent gravitational attraction on a photon (body) towards a reference body may be taken (duly modified for the constant of proportion) as a measure of its matter content or 'rest mass'. When the measure of apparent gravitational attraction on a body, near the earth's surface, is related to earth and converted into the gravitational units, it gives the 'weight' of the body.

Let there be two photons 'A' and 'B' in space and they do not have common disc planes. Consider that the photon 'B' is in the disc plane of photon 'A'. Photon 'B' is in a different phase compared to photon 'A'. Gravitational efforts on photon 'A' in its disc plane, from the direction of photon 'B' and from the opposite direction are different. Since the photon 'B' breaks the continuity of 2D energy fields in the disc plane of photon 'A', gravitational action from that direction on photon 'A' is less than the magnitude of gravitational action on it from the unbroken 2D energy fields in the opposite direction. Excess gravitational action may produce inertial action on photon 'A' to move it towards photon 'B'. This acts like a partial apparent attraction on photon 'A'. (Apparent gravitational attraction is a combination of gravitational actions on both the bodies in consideration). In this case, there is no mutual apparent attraction between the photons. At the same time, photon 'A' tends to move towards photon 'B' but photon 'B' has no such action on it. Hence, the inertial action on photon 'A' does not qualify as apparent gravitational attraction. Magnitude of inertial action on photon 'A' depends on its matter content. Photon 'B' has no direct involvement in the inertial action on photon 'A', other than that the matter body of photon 'B' happens to be in the disc plane of photon 'A'.

Other than in special circumstances, it is not possible for two photons, moving freely in space, to have apparent gravitational attraction between them for more than an instant. This is because; both the photons are free spinning bodies moving at constant linear velocity. It is extremely rare for them, unless by deliberate means, to coincide in frequency, phase, polarity, direction of motion and exist in the same plane for more than an instant. Once they come in the same plane, their median planes may never coincide with each other again.

When the photons are constituents of 3D macro bodies, they are controlled and deliberately made to assume states of motion, in which their median planes coincide with each other frequently. Since there are millions of photons crammed into the small space of a 3D macro body and at the same time they need to maintain their inherent motions at critical speeds; there are many occasions for their median planes to coincide with each other in the same planes. In case of the photons, measure of average magnitude of apparent gravitational attraction (weight) has a direct relation to their frequency (thereby it has a relation to their rest masses – matter contents). Nevertheless, in cases of other 3D bodies, which have millions of photons in them, measure of apparent gravitational attraction (or weight) has only an approximate and average relation to their rest masses or matter contents of the macro bodies.

Gravitational attraction between coplanar photons:

Value of gravitational constant, determined experimentally for larger macro bodies, caters for the measure of average and approximate value of the apparent gravitational attraction between two 3D macro-bodies (between constituent photons of two macro bodies, which happens to be in the same plane at any given instant). Each photon of both macro bodies contributes its part to this apparent gravitational attraction. A photon in one macro body may come in the same plane with many photons in the other macro body. All such coincidences contribute towards apparent gravitational attraction. Therefore, the contribution of any one photon in a large macro body to the apparent gravitational attraction between it and another macro body depends on the total number of photons in the other macro body (its matter content – rest mass).

Apparent gravitational attraction between two photons, one in each body, develops only when their disc planes happen to be in the same plane. This may happen rarely and at the most, only twice during every spin of the photon. Time duration for these two photons to be in the same plane is extremely short. Therefore, in normal cases, when photons are constituents of macro-bodies, their contribution towards the apparent attraction between the macro bodies is extremely small. This is only because of the extremely short duration of their stay in any common plane. Consequently, when we have to determine the apparent gravitational attraction between two photons, whose disc planes are continuously in the same planes, gravitational constant (used for larger macro bodies) needs to be modified. The fact that these photons' disc planes are continuously in the same planes and they are permanently (instead of twice every spin) under the mutual apparent gravitational attraction should be taken into account.

To be continuously in the same planes, two photons of equal rest masses need to be moving in the same (linear or angular) direction with their axes in a straight line and spinning in phase with each other. They have to spin at the same angular velocity and the distance between them has to remain constant. To satisfy these conditions, these photons have to move either:

(1). In parallel paths, perpendicular to their axes, moving in the same linear direction and the photons spinning in phase with each other or

(2). In a circular path with their axes coinciding with the diameter of the path and the photons spinning in phase with each other.

Under these conditions, the apparent gravitational attraction between the two photons is continuously and fully effective in all the planes containing disc planes of both the photons.

A photon can never stay in the same plane for more than an instant. Consider a hypothetical photon with theoretical maximum matter content. Such a photon is a sphere of 3D matter and remains continuously in all planes passing through it. It has its disc plane in all the planes passing through the sphere. It can also be assumed that because of its continuous presence in all planes passing through it, such a photon spins at theoretical maximum frequency. (Practically, due to the lack of disc faces, such a photon cannot spin at all). This maximum frequency can be taken as a measure of total number of disc planes of the photon, containing its spin axis. Since this photon maintains its presence in all the planes, apparent gravitational attraction involving this photon with any other photon also is continuous in each (or all) of these planes, provided the disc planes of the other photon remain in the planes passing through the hypothetical photon. It provides for a constant magnitude of apparent gravitational attraction in all planes passing through both of them.

As a general consideration, it was taken that the frequency of a photon is proportional to its matter content. Theoretically, a photon of maximum matter content should be spinning at theoretical maximum possible frequency. A photon of maximum matter content in 3D space system exists, simultaneously, in as many planes as there are planes (containing its spin axis) of existence of a photon of frequency of one Hz. multiplied by its (maximum) frequency.

As for the apparent gravitational attraction in a plane between two photons, continuously in the same planes, the magnitude of apparent gravitational attraction between them in one plane is the same as that between two photons of maximum frequency, because there is only one plane passing through the disc planes of both the photons. These photons are under constant apparent gravitational attraction instead of twice every spin. Therefore, in an equation, to determine the apparent gravitational attraction between two photons (continuously in the same planes), gravitational constant 'G' has to be modified by multiplying it with a number equal to the theoretical maximum frequency, (f_{max}), of a 3D photon of maximum matter content.

Magnitude of apparent gravitational attraction between two photons of rest masses 'm_1' and 'm_2', continuously in the same planes and at a distance 'd' between them;

$$GF = \frac{G \times f_{max} \times m_1 \times m_2}{d^2} \qquad (5/4)$$

This is the actual theoretical value of the apparent gravitational attraction between two disturbances (photons) continuously in the same planes. This value is too large compared to the value obtained by using the present gravitational constant 'G', determined strictly for the use with larger macro bodies. Magnitude of this constant of proportion, 'G f_{max}', in each plane, works out to be about 2.2 x 10^{37} m^3 / Kg sec^2. Therefore, the gravitational effort is not a 'weak force' as regarded today but it is an enormously strong, compared to other manifestations of natural force. It appears to be very weak because only an extremely small part of the rest mass of a body contributes towards its by-product, the apparent gravitational attraction, at any instant. Presently, apparent gravitational attraction between matter-bodies is the only known effect of gravitation. Note that it is not correct to compare between various 'forces' based on their strengths. All forces are similar and they are only manifested in various forms. Hence, basically, there cannot be any difference in their strengths.

Because of the extremely high value of apparent gravitational attraction between two coplanar disturbances, small independent 2D disturbances, formed in a 2D energy field at different places in a plane, merge almost instantaneously.

Should the coplanar photons be of different matter contents

or they shift from common planes, certain thickness of the larger photon is left out of planes of existence of the other. Therefore, the lower of the two rest masses of these two photons, 'm', should be taken in the equation (5/4) in place of the rest masses of both the photons;

Thus, the apparent gravitational attraction between two photons;

$$GF = \frac{G \times f_{max} \times m^2}{d^2} \qquad (5/5)$$

Part of a larger photon's body is outside the planes of a small photon's body's existence. This part has curved periphery, but it exists in the planes intercepted by the flat faces of the smaller photon. Smaller photon experiences no partial apparent attraction towards the larger one. Larger photon experiences additional partial apparent gravitational attraction towards the smaller one. Moving effort, experienced by the smaller photon, is lesser and the moving effort, experienced by the larger photon, is greater.

Photons are 3D bodies. Each photon is acted upon by all 2D energy fields of its existence. A photon of higher matter content is thicker and hence exists in more number of 2D energy fields. Action of gravitational efforts is greater on a larger photon. Figure 5D, represents four sets of photons, each set has two photons, spinning about a common axis, YY. Only one segment of each photon is shown in the figure, looking from the side on the curved periphery of one body-disc segment. Sizes of the photons are greatly exaggerated. $X_p - X_p$ and $X_q - X_q$ are the equators of the photons.

In the set of photons, containing P_1 and Q_1, both photons are of equal matter content. They spin at the same speed and keep their bodies always in the same 2D energy fields. Their bodies break the continuity of all 2D energy fields in the common planes passing through their bodies. Both photons experience equal apparent attraction towards the other. Magnitude of apparent gravitational attraction between the photons is according to the equation (5/12). Moving efforts experienced by both the photons

Chapter 5. GRAVITATION IN 3D SPACE

are equal and proportional to their matter contents.

In the set of photons, containing P_2 and Q_2, photon P_2 contains more matter than photon Q_2. They spin on common axis YY. Body of photon P_2 occupies few more planes than the number of planes occupied by the body of photon Q_2. Apparent interaction between the photons in between planes AA and BB produce apparent attraction between the photons as per the equation (5/5). Matter content in photon P_2, in the segment CX_qA and segment BX_qD produce partial apparent attraction due to gravitation on photon P_2 towards the photon Q_2. Thus, the photon P_2 experiences additional moving effort towards photon Q_2 over and above the apparent

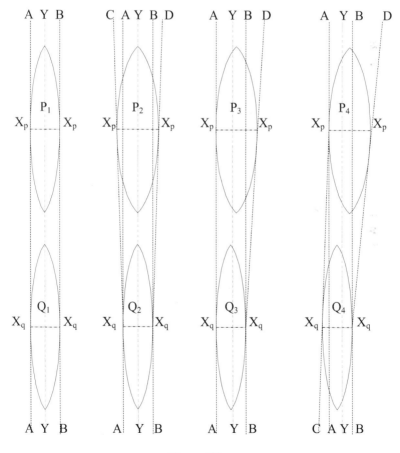

Figure 5D

gravitational attraction, it experiences towards Q_2 due to its matter content equal to the matter content of Q_2. Moving efforts experienced by the photons are unequal and are proportional to their matter contents.

Spin speed of a photon is proportional to its matter content. Hence, the photon P_3 in the set of photons P_3 and Q_3, (being larger) spins at a higher speed. As shown in the figure, photon P_3 is on the verge of overtaking the photon Q_3 in its angular displacement. Both photons are spinning about the common axis YY. Centre of peripheral segments of the photons, shown in the figure are displaced from the line YY. This is because the common spin axis of the photons, YY, passes through the centre of the photon-discs and the segments of the photons, shown in the figure, are at the curved periphery of the photons. Matter bodies of both photons intercept planes between AA and BB. Due to this intersection, photon P_3 and the photon Q_3 experience mutual apparent attraction due to gravitation as per equation (5/5). Due to excess matter content of photon P_3, within the segment BXD, it experiences additional partial apparent gravitational attraction towards the photon Q_3. Moving effort experienced by P_3 is greater than that experienced by the photon Q_3. Moving efforts experienced by the photons are unequal but are proportional to their matter contents.

In the set of photons P_4 and Q_4, photon P_4 is shown in advanced angular displacement with respect to the photon Q_4. Matter contents on the photons in common planes, shown between lines AA and BB are less than the matter contents of both photons. Part of photon P_4, between AA and BB experiences apparent gravitational attraction and part of the photon contained in the segment BX_qD experiences partial apparent gravitational attraction towards the photon Q_4. Similarly, Part of photon Q_4, between AA and BB experiences apparent gravitational attraction and part of the photon contained in the segment CX_pA experiences partial apparent gravitational attraction towards the photon P_4. Moving efforts, experienced by the photons remain, more or less constant until the angular displacement is too large. Magnitudes of the moving efforts are proportional to their matter contents.

In the special cases, where two photons of the same (or very nearly the same) magnitude of matter content and spinning about the same (or very nearly the same) axis, the apparent gravitational attraction between them is divided (among the photons) in proportion to their matter contents. Magnitudes of these apparent attractions remain continuous. In all other cases, apparent gravitational attraction between any two photons, whose median planes happen to be in the same plane, is of constant magnitude and lasts only for an instant, when their disc planes coincide.

5.1.4. Macro bodies' apparent gravitational attraction:

When 3D macro bodies of larger sizes are considered, gravitational interactions are between each of the disturbances (photons) in both the macro bodies (total matter content of all constituent photons of the bodies). A very large number of photons constitute a 3D macro body. Each constituent photon of a macro body produces apparent gravitational attraction with all constituent photons of the other macro body, as and when their median disc planes are in the same planes. Apparent gravitational attraction between two macro bodies is between their constituent photons, whose disc planes are in the same plane at a given instant.

To produce apparent gravitational attraction between two photons, it is necessary that median disc planes of both photons occupy the same plane and break the continuity of the 2D energy field in that plane. However, gravitation can act only on curved periphery of a photon (refer section 3.3.4). If two photons are out of phase with each other and their median disc planes are in different planes, apparent gravitational attraction is available only on one photon, which has its curved perimeter towards and away from the other photon. Even though the photons are continuously under gravitational actions, their straight perimeters cannot experience translation motion due to gravitation. If two photons, in the same plane, have their faces (straight perimeter) towards each other, they experience no apparent attraction between them due to gravitation. Generally, it may be taken that: when two photons have their median disc planes in the same 2D energy field, they experience full apparent attraction due to gravitation between them. When one photon intercepts the median disc plane of another photon in a 2D energy field, the photon, whose median disc plane is in that 2D energy field, experiences apparent gravitational attraction (magnitude of attraction reduced by half) towards the other. The other photon, whose median disc plane is not in the common 2D energy field, experiences no apparent gravitational attraction towards the other photon.

For general purposes, phase of the photon-discs and the angle subtended (mutually by the photons on each other) need not be taken into consideration. All photons are of the same radial size and photons in different macro bodies are too far from each other to make a difference in the angle subtended on each other. At any given instant, total apparent gravitational attraction between two macro bodies is constituted by:

(1). Full apparent gravitational attraction between photons, whose median disc planes happen to be in the same plane and in phase, with respect to each other and

(2). Partial apparent gravitational attraction between photons, whose matter cores happen to be in the same planes but their median disc planes out of phase with each other.

Let there be two macro bodies 'A' and 'B'. Apparent gravitational attraction is produced between a photon in body 'A' and another photon in body 'B', whenever their median disc planes come in the same plane. It is very rare for one photon in macro body 'A' to come in the same plane with another photon in macro body 'B', frequently. Hence, the contribution of a photon in any of the bodies, in conjunction with another photon in the other body, towards apparent gravitational attraction between the macro bodies is extremely small. Once median disc plane of a photon, in macro body 'A', comes in the same plane with median disc plane of another photon, in macro body 'B', they may come in the same plane after a very long time or not at all. During this time, these two photons together cannot contribute towards the apparent gravitational attraction between the macro bodies 'A' and 'B'. Each of these photons may subscribe towards apparent gravitational attraction by pairing with any other photons in the other macro body. However, apparent gravitational attraction between two macro bodies is felt continuously. This is because of the profuse number of photons in both the macro bodies, few of which are in the same planes at any given instant. At any instant, only an extremely small number of the constituent photons of both the macro bodies subscribe towards apparent gravitational attraction between the two macro bodies.

Size of a body in the 3D space system, for the purpose of apparent gravitational attraction, is represented by its matter content, in planes passing through the matter particles (median planes of the photons) of both the bodies. Total matter content of a body (disturbance) is represented by its rest mass. Relativistic mass, being only an apparent increase in the mass of the body due to its linear speed, cannot affect the gravitational actions on a body. Number of photons, in a macro body, subscribing to the apparent gravitational attraction (at any instant) can be taken as proportional to the body's rest mass. Each photon, in a macro body, has its apparent attraction with all other photons in the other macro body. Hence, the total number of apparent interactions is equal to the number of photons in one macro body multiplied by the number of photons in the other macro body. Taking the rest mass of a macro body equal to its matter content, magnitude of apparent gravitational attraction 'GF', between two macro bodies of rest masses m_1 and m_2, situated at a distance 'd' between their centers and all the constituent photons of both bodies being in the same planes;

$$GF = 2\left(\frac{1}{4\pi} \times \frac{G_2}{d^2}\right) m_1 m_2 = \frac{G_2}{2\pi} \times \frac{m_1 m_2}{d^2} \qquad (5/6)$$

When these macro bodies are large, each macro body contains millions of photons. Out of these photons, there are only very few photons in both the macro bodies that are in the same planes and subscribe towards the apparent gravitational attraction between the macro bodies, at a given instant. Number of photons in the macro bodies, which are in the same planes at a given instant, are an extremely small fraction of the total number of photons in them. Hence, the gravitational apparent attraction between two macro bodies, at any instant is related only to very small parts of their matter contents. In order to compensate for this, the constant of proportion in the equation, to determine magnitude of apparent gravitational attraction between two macro bodies, is determined from experience. A constant of proportion of value 'G', determined from practically measuring the apparent attraction between two reference bodies, is substituted (for $G_2 \div 2\pi$) in the above equation.

Thus, apparent gravitational attraction between two 3D macro bodies:

$$GF = \frac{m_1 m_2 G}{d^2} \qquad (5/7)$$

Apparent gravitational attraction, given by equation (5/7), is the total apparent attraction between two macro bodies. The term 'd', used in the equation represents the distances between the photons in the macro bodies. When considering the gravitational attraction between two macro bodies, it is not practicable to use distances between each pair of photons in them. Instead, we use the distance between the centers (of gravity) of the macro bodies. Using the distance between the centers (of gravity) of the bodies has its advantage and disadvantage. We use equation (5/7) to determine the inertial action of apparent attraction on one body without any consideration to other body. Unintentionally, the other body is assumed static in space and total inertial action is assigned to the body in consideration. For relative considerations, one of the bodies, usually, the larger one is considered static. It is like; a stretched rubber band is put around both the bodies. The rubber band pulls both bodies towards each other at speeds corresponding to their rest masses. If one body is held static, the other moves at greater speed towards the static body. Since no large body can be held static in space, this does not happen in case of gravitational apparent attraction. Both the bodies move towards each other. This gives a clear indication that the gravitational action is not mutually attractive but it acts on each of the bodies separately to move them towards each other by push actions.

Gravitational actions on each 3D matter particle in macro bodies are inherent about them. Presence of 3D matter particles of different matter contents or different orientations in the macro bodies cause variation in the strength of gravitational actions in different directions. Resultant actions, formed by gravitation, on different macro bodies appear as attraction between theses macro bodies. No new forces are generated and no force-carrying particles are transferred (or transmitted) between these macro bodies. All changes are modifications to the existing gravitational actions. No time delay is required for the modification of apparent attraction due to gravitation on changes in the parameters of the macro bodies. They appear instantly.

5.1.5. Magnitude of apparent gravitational attraction:

Figure 5C shows two 3D macro bodies; 'body I' has many photons of equal matter content, spinning about their axes in various planes. Photons 'R, S and T' have their axes in parallel planes. Photons 'Q and U' rotate about axes in different planes. 'Body II' has only one photon, 'A', of the same matter content as the other photons. Its axis of spin is in a plane, parallel to the planes containing spin axes of photons 'S, T and R'. Dotted circles represent the equators of the photons. Twin-triangles in the circles show the segments of photons' matter cores. Since all photons are of the same matter content, they spin at the same speed. We shall assume all of the photons spin in the same direction. During their

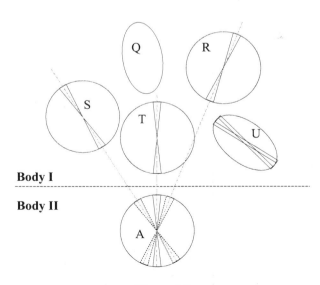

Figure 5C

rotation, photons 'R, S and T' in body I appear in the median plane of the photon 'A' in body II at different times (three times during every half-spin, as shown in the figure by the centre lines). Apparent gravitational attraction is active between body I and body II, whenever these photons come in a common plane. During every half-rotation of photon 'A', it aligns with photons 'R, T and S', once with each of them, in different angular positions. On each occasion of alignment, apparent attraction develops between the bodies.

As shown in the figure, both the bodies have three occasions to develop apparent gravitational attraction during every half-spin of the photons considered. Therefore, magnitudes of apparent gravitational attraction on both the bodies are equal. Actions on photons, whose median planes are blocked by plane surfaces of photons in the other body, are disregarded here. Difference in matter contents of individual photons plays its role only when we consider apparent gravitational attraction between two photons. In case of macro bodies, irrespective of their total matter content, number of photons in them or their sizes, apparent attraction applied on both the bodies are (approximately) equal. Minor differences due to photons, whose median planes are blocked by plane surfaces of photons in other body and due to differences in the matter contents of photons in the bodies, more or less balance their actions on both bodies.

All actions are recognized by inertial motions. Work, introduced in the 2D energy field of any plane, determines the magnitude of apparent gravitational attraction in that plane. Apparent gravitational attraction causes inertial motion of matter content in that plane.

However, when the bodies are three dimensional and large, there are many 2D energy fields passing through and common to both bodies. Depending on the magnitude of matter content (parts of matter contents in 3D matter particles with curved perimeter, in the plane) in the 2D energy fields considered, apparent gravitational attraction produced in it may differ from the apparent gravitational attraction, produced in other planes. Apparent gravitational attraction on a macro body is considered as a single force. For this we have to take the average value of apparent gravitational attractions in all the 2D energy fields passing through and common to both the bodies.

5.2. 3D system's gravitational constant in 2D system:

Let G be the constant of proportion in the equation for the magnitude of apparent gravitational attraction between two (spherical) macro bodies of mass, 'M' each, situated at a distance, 'd', between their centers of gravity, in 3D space system. In determining the apparent gravitational attraction in 3D space system, all matter contents in each of the bodies is assumed to be concentrated at their centers of gravity. Centre of gravity or centre of mass is a geometrical point of zero dimensions. Taking, 'd', as the distance between the centers of the bodies has the advantage of assuring a compensation factor in the equation. As long as the bodies have positive existence, the factor 'd' does not become equal to zero and take the result of the equation to infinity. In this case, 'd' can be equal to zero only in case the mass 'M' is also brought down to zero. Any (spherical) body that has mass 'M' (assuming the mass represents body's matter content) has to have a positive value of radius and thus make distance between the centers of two bodies, greater than zero. Thus;

Magnitude of apparent attraction due to gravitation between two bodies of equal matter content in 3D space system;

$$F = \frac{M^2 G}{d^2} \qquad \text{By equation (5/7)}$$

Constant of proportion of apparent gravitational attraction;

$$G = \frac{Fd^2}{M^2} \qquad (5/8)$$

When, mass 'M' and distance 'd' are equal to unit measures, gravitational constant, G is equal to the magnitude of apparent attraction, F. For this reason, size or shape of the concerned 3D macro bodies do not come into consideration to determine gravitational constant, G, in 3D space system, except as a compensating factor in the equation. For the current gravitational equations to be true there can be only one centre of gravitation in a body.

Assuming a single body to have different centers of gravitation for different parts of the body is distorting the theory. The very definition of centre of gravity requires that gravitational actions on all parts of the body to subscribe towards the average location of their actions. It is not right to assume the same apparent gravitational attraction, on a macro body, can affect different parts of the body at different magnitudes. One of the primary assumptions in the above equation is that the whole matter content (mass) of a body is concentrated at its centre of mass, which often acts as the centre of gravity also. No body can have more than one centre of gravity. [This incorrect logic is currently used to explain the mechanism of terrestrial tides].

Magnitude of a disturbance is the extent of its contact with 2D energy fields. All 3D matter particles are disturbances in the 2D energy fields. Matter content in each plane is in contact with 2D energy field of the same plane. For a 3D matter particle, extent of its contact with 2D energy fields equals its surface area. A 3D macro body is a union of numerous 3D particles (photons). In a 3D macro body, basic 3D matter particles are situated far apart from each other and inter-particle field forces hold them together. This makes the matter-density of a 3D composite macro body lesser than the matter-density of a 3D matter particle. Matter-densities of all basic 3D matter particles (photons) are the same. It is their distribution in a macro body, which makes the matter-density of a composite macro body, lesser and different. A macro body, with its matter particles distributed farther from each other, has low matter density. A composite 3D macro body may be considered as single 3D disturbance of lower matter-density.

$$\begin{bmatrix} \text{Matter density of a 3D body,} \\ \text{considered as a 3D disturbance} \end{bmatrix} = \frac{\text{Total matter content of the 3D body}}{\text{Surface area of the body}}$$

$$\text{Magnitude of a 3D disturbance} = \begin{bmatrix} \text{Matter density of the disturbance} \\ \times \text{Surface area of the body} \end{bmatrix}$$

$$= \frac{\left[\begin{array}{c}\text{Total matter content of the 3D body}\\ \times \text{Surface area of the body}\end{array}\right]}{\text{Surface area of the body}}$$

$$= \text{Total matter content of the 3D body}$$

$$= \text{Rest mass of the body.}$$

Matter content of a 3D disturbance, equivalent to a 3D macro body, is the total matter content of the 3D body. Total matter content of a body is represented by its rest mass. Mass of a body is used to evaluate the apparent gravitational attraction between 3D bodies. Assumption that the whole of matter content of a macro body is concentrated at its centre of gravity is required, to have a common point to represent the matter content of the macro body. However, a 3D disturbance or its equivalent, as considered above, is a single body and it has to have a continuous surface area. Same macro body cannot be considered as number of smaller parts and have different or fractional gravitational actions, separately for each part.

For us, the 3D beings, matter bodies have existence only when they exist in all three spatial dimensions of the 3D space system. It should have a volume. Since a plane has no thickness, a 2D body in three-dimensional system has no existence (volume) in 3D space system and its matter content and mass becomes zero in calculations. However, it may be understood that a 3D body has its existence in every 2D plane passing through it. Matter content of such a body is distributed in each of these planes, proportionately. Volume or matter content of a 3D body is the sum of numerous parts of the body in as many planes. Since, at present we have no practical two-dimensional space system, we may use the present value of the gravitational constant for 2D bodies, with corresponding modifications to the units involved. A plane of unit thickness that can be used by present equation is a 'meter' (being one unit of distance) thick. This may not influence calculations of apparent gravitational attraction, in 2D space system, between very large bodies at great distances. However, if the thickness of a plane in 3D space system is reduced to a smaller unit (less than one meter), corresponding changes in the gravitational constant is also required. If the unit measure of distance is reduced to 0.5 meter, corresponding change in the value of gravitational constant is four times enhancement.

Additional work, introduced into the matter field of a body in any particular direction is contained only in the 2D energy fields in that particular direction. Figure 5D shows two spherical macro bodies, 'A' and 'B' in free space. Body 'A' is very large. Body 'B' is very small and situated very far from body 'A'. We shall consider only those actions specifically subscribing to apparent gravitational attraction between these bodies. Extents of 2D energy fields, contributing to apparent gravitational attraction on body 'A' on the outer side are those included in the conical sector CDQR and passing through both bodies. Similarly, extents of 2D energy fields, contributing to apparent gravitational attraction on body 'B' on the outer side are those included in the conical sector FGMP and passing through both bodies. On the inner side of the bodies, extents

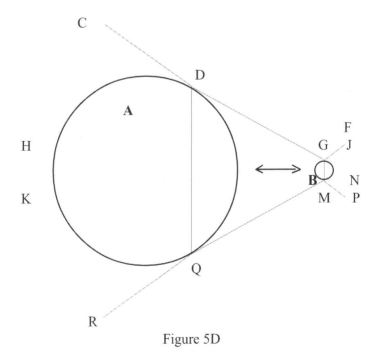

Figure 5D

of 2D energy fields contributing to apparent gravitational attraction between the bodies are those enclosed in the conical sector DGMQ and passing through both bodies. These 2D energy fields contribute towards overall apparent gravitational attraction between the bodies. Average of all apparent gravitational attractions in all these 2D energy fields determines the total apparent gravitational attraction between the bodies.

Out of all these 2D energy fields, contributing to the total apparent gravitational attraction between these bodies, those 2D energy fields in planes passing through both bodies and contained in the tubular sector HJKN, contribite towards apparent gravitational attraction between these bodies in the direction of the double headed arrow. Instantaneous inertial movements of the bodies towards each other in the direction of double headed arrow are determined only by the average magnitude of work introduced into their matter fields by these 2D energy fields. However, depending on the magnitude of matter contents in each of these 2D energy fields, work introduced in each of them may differ from magnitude of work introduced in others. Hence, we must consider the apparent gravitational attraction in each plane different from apparent gravitational attraction in other planes.

Each plane passing through both bodies has its own contribution to the total apparent attraction due to gravitation between the bodies. By adding the efforts in all planes may give us the total apparent gravitational attraction between the bodies. However, inertial actions produced in each plane correspond to the magnitude of apparent gravitational attraction in each of the planes. Hence, the total inertial action on the whole body corresponding to the total magnitude of apparent gravitational attraction between the bodies corresponds to the average magnitude of apparent gravitational attraction available in each plane.

Directional apparent gravitational attraction between two 2D bodies may be calculated by using the gravitational constant in 3D system as follows:

Substituting the 3D space system's gravitational constant 'G' for '($G_1 \div \pi$)' in equation (3/4);

Apparent gravitational attraction between two 2D bodies;

$$F = \frac{m_1 m_2 G}{d} \tag{5/9}$$

Where m_1 and m_2 are rest masses of two 2D bodies, 'd' is the average distance between points on their peripheries and G is the gravitational constant used in three-dimensional space system. We shall estimate the apparent gravitational attraction between a 2D matter particle, of unit rest mass, situated in the equatorial plane of a small spherical body and the matter content of the equatorial plane of another very large spherical body, in the same plane. Only the matter contents in the same plane can provide apparent gravitational attraction.

Let 'R' be the radius of the large spherical body, 'M' its mass and 'd' the average distance between the peripheries of the 2D body and the very large spherical body.

$$\text{Volume of a large spherical body} = \tfrac{4}{3}\pi R^3$$

$$\text{Matter density of the large spherical body} = \frac{M}{\tfrac{4}{3}\pi R^3}$$

$$\text{Area of equatorial plane of the large spherical body} = \pi R^2$$

Mass of equatorial plane of unit thickness = Matter density × Area

$$= \frac{M}{\tfrac{4}{3}\pi R^3} \times \pi R^2 = \frac{3M}{4R}$$

$$\begin{bmatrix} \text{Matter density of equatorial plane} \\ \text{of the large spherical body,} \\ \text{considered as a 2D disturbance} \end{bmatrix} = \frac{\text{Mass of equatorial plane}}{\text{Surface area of equatorial plane}}$$

$$\text{Matter density of the 2D disturbance} = \frac{\text{Mass of equatorial plane}}{\text{Surface area of equatorial plane}}$$

$$= \frac{3M}{4R} \div \pi R^2 = \frac{3M}{\pi 4R^3} \text{ kg/m}^2 \quad (5/10)$$

Equatorial plane has its contact with 2D energy field at its perimeter.

Perimeter of the spherical body's equatorial plane = $2 \pi R$

Magnitude of 2D disturbance = $\begin{bmatrix} \text{Matter density of 2D disturbance} \\ \times \text{Perimeter of the disturbance} \end{bmatrix}$

$$= \frac{3M}{\pi 4R^3} \times 2\pi \times R = \frac{3M}{2R^2} \quad (5/11)$$

Magnitude of apparent gravitational attraction between matter content in the equatorial plane of the spherical body and a 2D body of unit mass, in the same plane;

$$F = \frac{3M}{2R^2} \times \frac{G}{d} = \frac{3MG}{2dR^2} \quad (5/12)$$

Where, 'd' is the average distance between points on their perimeters.

This apparent attraction acts between the equatorial plane of the large spherical body and a 2D body of unit mass in the same plane on the surface of a small spherical body. Similar apparent attractions in every plane of 3D space, containing two bodies, subscribe to the total magnitude of apparent gravitational attraction between them in 3D space system. When, taken together, these separate apparent attractions add to give the apparent attraction between the two whole bodies in 3D space system. Action in each plane is distinct and it takes place only in the common planes, containing both the bodies. If certain part of one body does not have a common plane with the other body, no gravitational action is produced between that part and the other body.

Action of central force between earth & moon and earth & sun are directional. Directional apparent gravitational attraction between bodies in each pair may be determined by equation (5/

12), for each plane of earth, perpendicular to its spin axis. Magnitude of apparent attraction between earth and moon (experienced at every plane of earth, parallel to moon's orbital plane about the earth) is about 2.3 times greater than that between earth and sun (experienced at every plane of earth, parallel to sun's orbital plane about the earth). This ratio is same as the ratio between lunar tides and solar tides on earth. This gives a logical reason why the magnitudes of lunar tides are larger in magnitude, compared to the magnitudes of solar tides.

5.2.1. Practical gravitational constant:

The constant of gravitation, in the equation for apparent gravitational attraction (between two macro bodies in three-dimensional system), has been determined by three methods:

(1) By comparison of the apparent pull of a large natural (celestial) macro body with the Earth,

(2) By measuring the apparent attraction of the Earth upon a test macro body with a laboratory balance, and

(3) By direct measurement of the apparent attraction between two macro bodies in the laboratory.

Magnitude of apparent gravitational attraction between two bodies in 3D space is proportional to the product of their rest masses and inversely proportional to the square of the distance between their centers of gravity. Practical constant of proportionality, 'G', is experimentally established to have a value: $G = 6.668 \times 10^{-11}$ m^3 / Kg sec^2. This value of G is extremely small compared to actual value of G_2 as given in the above equations. Value of G is experimentally determined by measuring the apparent gravitational attraction between two bodies of known rest masses, in 3D space system. These bodies consist of millions of photons, each spinning and moving in various directions at the same time. Chances for median disc planes of any two photons in these bodies (one in each body), to be in the same 2D energy field plane for more than an instant, are extremely small. Only because of very large number of photons in these bodies, we are able to get a continuous apparent gravitational attraction between the bodies.

Chapter 5. GRAVITATION IN 3D SPACE

Apparent gravitational attraction, between two macro bodies (at any instant), is actually contributed by very small number of photons in these bodies. Hence, the apparent gravitational attraction between the two 3D macro bodies, measured by us, corresponds to only a very small part of the total matter content of the macro bodies. Value of the constant of proportion of apparent gravitational attraction, G, determined experimentally, is based on the action of this small fraction of the matter content in each body. Since it corresponds to only a very small part of the actual matter content of the macro bodies, it is extremely small. This is how we came to consider the gravitational efforts, incorrectly, as a weak force in nature.

Value of gravitational constant, G, was determined purely on empirical parameters. It was determined without understanding the mechanism of gravitation. It cannot be derived mathematically from other known parameters. Contemporary knowledge of gravitation is based on many illogical assumptions and interpretations. Since the present value of gravitational constant, G, was determined without understanding the nature of gravitation, it is entirely based on human ingenuity rather than on scientific facts.

Actually, the theoretical value of gravitational constant 'G_2' (as given in this text) is so huge that the gravitational actions are able to account for the compression of functional quanta of matter into three-dimensional matter of density of about 10^{142} Kg/m³, in the photons and much more into the next dimensional space system. Theoretical value of 'G_2' in 3D space for the 3D matter (for matter bodies with physical parameters of photons) is about 2.2×10^{37} m³/Kg sec² (about 3.3×10^{47} times of its present value. Compare this with the present value of $G = 6.668 \times 10^{-11}$ m³/Kg sec²). Gravitational actions are enormously strong. Given value of G_2 is very approximate and is for a hypothetical case where all the photons in the interacting macro bodies are in the same 2D energy field plane. Hence, the present value of gravitational constant, determined for larger macro bodies is very inappropriate, when dealing with extremely small bodies (fundamental particles, subatomic particles, etc.), which are constituted by very few photons.

Gravitational action on a 3D macro body, is the resultant of all gravitational actions, separately on each of its photons. Apparent gravitational attraction between two macro bodies is the resultant of the apparent attractions due to gravitation between photons in both the bodies, in each of the planes, passing through both the bodies. Gravitational action in a plane depends on the curvature of the perimeter of a disturbance. Hence the gravitational action in the 3D space system, will also depend on the curvature of the surface of the photon-body (here a body refers to the basic matter particles of a macro body and not to the macro body constituted by the photons). All 3D macro matter bodies are constituted by photons and there are no stable 3D matter particles larger (in radial size) or different from a photon. All photons have the same radial size and have equal curvature of their disc circumference. As a result, curvature of the macro-body surface has no practical effect when considering apparent gravitational attraction between bodies, larger than subatomic particles.

Body surface, here, means the surface of independent basic 3D matter particles rather than the surface of a composite macro body, constituted by millions of photons spaced far apart and moving independently within the macro body. Since the photons are disc-shaped, their curved periphery is convex and their faces are (almost) flat. All photons exhibit equal (distance being equal) apparent gravitational attraction between them, in each of their disc planes and none in the other planes. When the median plane of one photon is blocked by disc face of another photon, only the photon with its curved perimeter (towards the other photon) experiences the apparent attraction towards the other. Sizes of all photons are uniform (other than for their matter content, which may have slight variation in different macro bodies). Whatever their matter contents may be, all photons take same time to spin through any one plane. Matter contents of photons, constituting different 3D macro bodies, vary within a very small margin only. So the number of photons contained in a larger macro body may be taken as proportional to the matter content (rest mass) of the body.

In this book, the gravitational action and the apparent gravitational attraction are distinctly separate. Gravitational action is a 'real force' exerted by the 2D energy fields on a disturbance in them. Apparent gravitational attraction is the resultant of gravitational actions on more than one body that may produce bodies' inertial motion. Separate inertial actions, produced on two bodies towards each other, appears to be attraction between them, resulting from the action of gravitational actions on the bodies. Hence, it is termed as 'apparent gravitational attraction'. If no inertial actions are noticed on the bodies towards each other, apparent gravitational attraction between the bodies is not envisaged.

At present, it is believed that the apparent gravitational attraction is the result of an interaction between two matter bodies at a distance and depends directly on their masses. Masses of these bodies depend on their speed. Therefore, the apparent gravitational attraction between two bodies, moving at very high speed, (keeping the distance between them constant) also should rise to high values. This does not happen. Therefore, it is believed that the apparent gravitational attraction between bodies is only due to their rest masses (3D matter contents). 2D energy fields can act only on real matter content of bodies, represented by rest masses and not on an apparent increase in their masses due to an increase in bodies' speeds of motion.

5.3. Action at a distance:

Matter is inert. It is the 2D energy fields, which produce and transfer the distortions in their latticework structure that act on the matter particles during exchange of work between two matter fields of different bodies. All actions are between each of the basic 3D matter particles in the bodies and the surrounding 2D energy fields. Matter bodies cannot act on each other even if they are in contact (feeble apparent attraction due to self-adhesive nature of matter in the quanta of matter is ignored, here). Only interaction possible between two 3D matter particles, in contact, may be a merger between them – combining their constituent quanta of matter to form a single matter body. Even for this to happen, actions of 2D energy fields on these bodies are essential.

Each matter particle produces distortions in the 2D energy fields. These distortions are transferred, in turn, through the latticework of the 2D energy fields to the distortion fields of any other matter particle. Transfer of distortions by each 2D energy field is limited to its own plane. Transferred distortions, in the 2D energy fields, act on any other matter particle to produce the (apparent) interaction between the matter particles. Consequently, there is no action at a distance. Actions of all efforts on matter bodies are similar. Thus, this concept does away with the myth of 'action at a distance' by various natural forces. There are no real interactions between two seemingly interacting matter bodies. Interactions are between the matter particles of each of the bodies and the 2D energy fields about them. Matter particles of the (seemingly) interacting bodies are directly in contact with the quanta of matter (latticework junctions) in the 2D energy fields. It is these quanta of matter of 2D energy fields, which are applying gravitation directly and separately onto each of the basic 3D matter particles. Hence, there is no interaction between two bodies during apparent gravitational attraction between them. Interactions are between each of the bodies and the 2D energy fields of their existence, separately.

A change in the parameter of a matter body makes instantaneous changes in the magnitude of its interaction with the 2D energy fields. This also makes instantaneous changes in the apparent gravitational attraction between the changing matter body and any other matter body in space. It is the interaction between the changing matter body and the 2D energy fields, which is modified and not the apparent interaction between the bodies considered. Hence, there is no transfer of forces or particles from one matter body to another, during variations of parameters of a matter body, to affect the magnitude of apparent gravitational attraction between them. Thus, (however far apart the bodies may be) it requires no time interval for a variation in the magnitude of apparent gravitational attraction between two matter bodies to develop. Any change is instantaneous.

5.4. Screening the gravitation:

All 2D and 3D matter particles are disturbances with respect to the 2D energy fields. A 3D macro body mainly consists of (vacant) space, filled with 2D energy fields – the matter field of the body – and their constituent quanta of matter. Actual 3D disturbances within the macro body – its matter content in the form of photons – are relatively very few. They are spaced far apart and are under constant motions within the macro body. Interactions between 2D energy fields and basic 3D matter particles of the macro body keeps the constituent photons moving at their critical speed in curved paths and spin them at spin speeds proportional to their matter contents. It is the slight changes in these photons' curved paths, which produces all actions, we notice.

Because of large space between the basic 3D matter particles in a macro body, most of the 2D energy fields pass through the macro body without being intercepted by the 3D disturbances (constituent photons) in the macro body. It is only those 2D energy fields, which are blocked (discontinued) with the convex curved periphery of the photons, which subscribe to effects of apparent gravitational attraction. Number of such photons is very few compared to total number of photons in the macro body.

The very structure of a 3D macro body has to have vast intervening distances between its basic 3D matter particles. Regions of space between these particles in the macro body are filled with 2D energy fields, in the planes passing through the body. Hence, it is the nature of 3D macro bodies to be extremely porous to the 2D energy fields passing through them. Because of this porous nature of the 3D macro bodies, it is impossible to screen or stop the 2D energy fields, completely from acting in any region, by using another 3D macro body. This is why; the apparent gravitational attraction exists between every 3D matter object in the universe – space – however far they may be, irrespective of the presence of other macro bodies of any density or shape, which happen to be in between them. A 3D macro body provides unhindered continuity to most of the 2D energy fields passing through it. This is also the reason why the magnitude of apparent gravitational attraction is a function of magnitude of rest mass of bodies (density × volume) rather than their areas or volumes.

To screen or stop the apparent gravitational attraction, it is necessary for a screening body to completely block all the 2D energy fields passing through it. Theoretically, it may be possible to screen the 2D energy fields and thereby remove apparent gravitational attraction between two bodies by intervention of a 3D macro body made of pure matter placed in between two bodies. Matter density of this intervening body will be approximately to the order of 10^{142} kg/m^3, which is the matter density of the photon. Even in this case, other 3D matter bodies are (apparently) attracted to this screening body, due to gravitation. In the 3D space system, it is impossible to make a body of pure matter larger than the size of a photon. There are no other means to screen a body from its gravitational effects, giving rise to the apparent gravitational attraction towards other matter bodies.

Currently, the apparent gravitational attraction is assigned directly to the matter bodies. Hence, it is assumed that the apparent attraction between bodies contributes towards determining their relative positions. Let two bodies be held in certain relative position, with respect to each other, by the apparent gravitational attraction acting in conjunction with other forms of natural forces. Since the bodies are held in stable state by an attraction, this effort should be acting against an equal magnitude of repulsion. Removal or screening of apparent gravitational attraction should move the bodies away from each other by those efforts, which were applied against the apparent gravitational attraction to keep the body in steady relative positions. Even if we are able to screen the apparent gravitational attraction between two bodies, this will not happen. Gravitational actions are between a body and the 2D energy fields. Therefore, placing a screen in between two macro bodies does not affect the separate interactions between the 2D energy fields and each of the two bodies. Apparent gravitational attraction depends only on the matter content of the body and the extents of 2D energy fields about them. Presence or absence of other bodies in the neighborhood does not affect the gravitational efforts, applied on a body.

5.4.1. Levitation:

Levitation is a phenomenon of raising a small macro body

Chapter 5. GRAVITATION IN 3D SPACE

from the surface of a very large macro body, with which the small macro body is held by the apparent gravitational attraction between them. Since there are no practical ways to screen a body from gravitational effects, in order to produce levitation, certain external effort has to be applied on the small macro body (preferably from in-between the bodies) to counteract apparent gravitational attraction between the macro bodies. If the external effort is sufficient to counteract the apparent gravitational attraction, the small macro body will float above the surface of the large macro body.

It is the differences in the gravitational efforts on either side of a macro body, which produces the apparent attraction due to the gravitation. Hence, a modification of the distortions in the 2D energy fields in between the bodies may increase or decrease the apparent gravitational attraction between the bodies. This could be achieved by appropriate 2D energy field-distortions introduced into the 2D energy fields, in between the macro bodies, from external sources. Introduction of linear distortion fields in between the macro bodies in the direction of the smaller macro body can increase the reaction on the small macro body from the space in between the macro bodies. Gravitation on the outer side of the small macro body remains the same. Increase in the reaction from the space in between the macro bodies can neutralize, partially or completely, the gravitational effort from the outer side. Apparently, the small macro body will no more be under (whole) apparent gravitational attraction towards the larger macro body. State of the small macro body, with respect to the larger macro body is determined by the relative magnitude of reaction.

If the gravitational effect on the outer side of the small macro body is partially neutralized, it may appear to lose weight. If it is fully neutralized by the reaction, the small macro body will appear to float in space. Unless external efforts affect the small macro body, distance between the floating body and the larger macro body does not vary. Even if the larger macro body is spinning, like earth, the apparent centrifugal force due to the rotary motion does not affect the distance between the macro bodies. Distortions in the 2D energy fields, near the surface of the large spinning body, travels with the surface of the large macro body. Smaller macro body is carried within these distortions and has no relative displacement with respect to the distortions. Hence, the smaller macro body will maintain its relative distance from a fixed point on the surface of the larger macro body. This phenomenon is the 'levitation' of the body.

If the reaction from the space in between the macro bodies is more than the gravitational effort on the outer side of the small macro body, the small macro body is pushed away from the larger macro body.

* ** *** ** *

Chapter SIX
MAGNITUDE OF ATTRACTION

Essence of Newton's theory of gravitation is that the force (of attraction due to gravitation) between two bodies is proportional to the product of their masses and the inverse square of their separation (between their centers of gravity) and that this force depends on nothing else. With a small geometric modification, the same is true in general relativity theory. During the latter part of the 19th century, many experiments showed the apparent gravitational attraction is independent of temperature, electromagnetic fields, shielding by other matter bodies, orientation of crystal axes, and many other factors. In this concept, attraction due to gravitation between two matter bodies is apparent in nature and it is the resultant of simultaneous and separate gravitational actions on the bodies.

6.1. Centre of gravity:

A force is assumed to act at point-bodies or through dimensionless points in larger bodies. When action of a force through a larger area is considered, more appropriate term, 'pressure' is used. The assumption that a force acts through points is far from reality and physically impossible. However, it is very convenient for analytical purposes. This is illustrated clearly in the case of gravitational actions.

Imaginary point in (or outside) a body of matter, where, for convenience in certain calculations, the total weight of the body may be thought to be concentrated is called the centre of gravity of the body. Although the term 'centre of gravity' is widely used, the same imaginary location in (or outside) a body may also be called body's 'centre of mass'. Because the centre of mass does not require the assumption of 'gravitational field' about a body, many physicists prefer the term 'centre of mass' to 'centre of gravity'. When a body is situated in a 'gravitational field', its centre of mass and centre of gravity share a common location. An exception is a pair of large cosmic bodies that exert apparent gravitational attraction on each other as each of them orbit around (an assumption) the other. In binary star systems, for example, the stars' mutual apparent gravitational attraction may cause a separation of the centre of mass from the centre of gravity in each of the bodies.

Location of a body's centre of gravity may coincide with the geometrical centre of the body, especially in a symmetrically shaped objects composed of homogenous material. An asymmetrical object composed of a variety of materials with different masses, however, is likely to have a centre of gravity located at some distance from its geometrical centre. In some cases, such as hollow bodies or irregularly shaped objects, the centre of gravity (or centre of mass) may occur in space at a point external to the physical material; e.g., in the centre of a tennis ball or a point between the legs of a chair. Action of an effort on a body, through a point outside body-material is a physical impossibility. The case about the centre of mass of a system of many bodies is also similar. Although these assumptions are able to give us correct results in mathematical analyses, physical aspects of such actions, described by these assumptions, cannot be real.

Every 3D matter particle (every point on its surface) in a macro body contributes towards the gravitational effects about the body. Basic 3D matter particles of a macro body are distributed throughout the volumetric space, occupied by the macro body. Hence, in order to get a single point of action for all gravitational actions on every matter particle of the body, a point is imagined, where the whole weight of the macro body is assumed to concentrate. Weight of a macro body denotes its acceleration due to gravitation (towards a much larger macro body), in appropriate gravitational units. This imaginary point is the 'centre of gravity' of the macro body. Since, weight and rest mass of a macro body (of uniform density) are proportional to each other, this point is also called 'centre of mass' of the macro body. Other than in certain special cases, centre of gravity and centre of mass of a macro body occupy the same location. This gives the convenience that size and shape of the macro body becomes immaterial in any analytical operations for apparent gravitational interactions between the bodies.

Originally, the centre of gravity was created only for symmetrical spherical bodies. Since the centre of gravity of a macro body is an assumed point in space, (although it is not very accurate) it could be anywhere in space with respect to a body of different shape. Only condition being that the resultant of all actions due to apparent gravitational attraction (through every point in the volumetric space of the macro body) should represent a single action through the centre of gravity of the macro body. Hence, the centre of gravity needs not be within the body. It needs to have no particular relationship to the shape or size of the body. Depending on the shape of a macro body, its centre of gravity either can be inside or outside the volumetric space occupied by the macro body.

Combinations of two bodies, where one body is placed within the other and a single continuous surface (of the outer body) enclose both bodies; both of them together form a single body. It will have only one centre of gravity. Location of this centre of gravity will depend on the relative position of the bodies and their matter distribution. There could also be odd shaped bodies, which when placed in certain configuration behave like a single body for purposes of (apparent) attraction due to gravitation. Though there are millions of matter particles in a macro body, the body is considered as a single unit. All basic 3D matter particles in a macro body together form an integral unit by mutual (apparent) gravitational and field forces. This composite body may be considered as a single body. Individuality of a single body is exhibited by its closed outer surface. This macro body has a single matter field. Matter field of a body, rather than its constituent particles, represents the body in space. Whatever the shape of a single body is, its outer surface is continuous and complete. That is its matter field is a wholesome single unit in the 2D energy fields. Hence, a single body cannot have more than one centre of gravity or different centers of gravity for body's different body-parts. In most of present explanations on the cause of tides on earth, different parts of earth are assumed to have different centers of gravity. This is corrupting the phenomenon of centre of gravity.

Different bodies of a system, which are mechanically linked, have a single continuous outer surface. Hence, they have a common centre of gravity. This system can not have separate centers of gravity for its constituent individual bodies. Different bodies of a system, which are not mechanically linked, cannot be considered as a single body. However, for the purpose of calculations they may have common centre of mass but they do not have a common centre of gravity. Each body of the system will have its own separate centre of gravity. Envisaging common centre of gravity for the bodies of a system of mechanically unlinked bodies, similar to common centre of mass in some theories on planetary motions, is stretching the phenomenon of centre of gravity too far.

Consider apparent gravitational attraction, felt by a very large macro body towards another external body. Magnitude of apparent gravitational attraction felt by basic 3D matter particles nearer to the external body are greater than similar attraction felt by basic 3D matter particles farther from the external body. This can shift the centre of gravity of the very large macro body towards the external body.

Consider the magnitude of apparent gravitational attraction by a large (moving) macro body towards an external static body (or that has a relative motion with respect to the moving macro body), in any particular direction. Action in any direction considered depends on the magnitude of work introduced by the apparent gravitational attraction in that direction. For a constant magnitude of apparent gravitational attraction, magnitude of work introduced in any direction depends on the duration, when planes in both the bodies (in the direction considered) coincide.

As the moving macro body overtakes the external body, planes in the rearward part of the macro body will be in line with the external body for longer time. Hence, forward part of the macro body feels smaller apparent gravitational attraction compared to rearward planes of the body. As a result, centre of gravity of the macro body shifts towards the rear. Magnitude of shift is proportional to their relative speed. This phenomenon, during the action of central force, shifts the centre of gravity of a planetary body rearward. [See paragraph 7.3.] Displacement of centre of gravity from its centre of mass reduces linear component of central force and creates a torque to spin the planetary body, about its centre of mass.

6.1.1. Inverse square law:

Equation (5/5), $GF = (m_1 m_2 G)/d^2$, is used to determine the magnitude of apparent attraction due to gravitation between 3D macro bodies. By the natures of its factors, this equation is named 'inverse square law of gravitation'. Generally, this equation is considered a 'universal law'. That is, it is true for all values of its variants. Inverse square law of gravitational attraction is assumed to determine the magnitude of (apparent) gravitational attraction between two macro bodies, external to each other. For any value of bodies' parameters, the equation is assumed to hold good. [Because of its universality, the same law is adapted for calculations of all natural forces, with appropriate substitution of relevant changes in the functions]. Although this law is considered as universally true, it has its drawbacks, which are always overlooked. Equation, used in the law of universal gravitational attraction (or the inverse square law of gravitation) gives us the best approximation of the magnitude of (apparent) gravitational attraction between two rest masses (of macro bodies). It gives us the mathematical relation between different derived parameters of the bodies, in terms of magnitude of apparent gravitational attraction, between their rest masses.

This equation does not rely on macro bodies' real parameters, contributing to the (apparent) attraction. Mass, used in the equation to represent matter contents of the bodies, is a mathematical relation between body's matter content and the body's acceleration under action of certain external force. In some cases, magnitudes of bodies' masses, considered for the equation are limited by their relative locations.

Distance between the bodies, d, used in the equation, is the distance between their assumed 'centers of gravity'. This is, subject to certain conditions, which depends on the relative locations, sizes and shapes of the bodies.

In static conditions, matter content of a macro body is related to its matter density and volume. In case of spherical bodies (and in case of most other bodies of solid shapes), a macro body of positive existence has certain volume and corresponding radius.

This ensures that however small two macro bodies are, (even if the bodies are in contact), there is a positive distance between their centers of gravity. Having a positive distance between centers of gravity of bodies gives a compensating factor, which prevents the term 'distance' in the equation from becoming or approaching zero magnitude. Reducing the distance between centers of gravity of two bodies also requires corresponding reduction in its size and hence its matter content (mass). In cases of regular shaped macro bodies, centers of gravity of two bodies can touch only if their masses are zero as well. Practically, this is an impossible proposition. Reducing the distance between centers of gravity of two bodies towards zero magnitude will increase the result of the equation (inverse square law) towards infinity, which is prevented by the compensating factor.

There may be bodies of odd shapes, whose centers of gravity could coincide in space within or outside their borders. In such cases, if their full (rest) masses are taken into consideration, equation of inverse square law will fail to provide any viable result. Since the equation is applicable to external bodies, it is not suitable to find magnitude of gravitational attraction between bodies, of which, one body may be situated (partially or fully) within the other. This is rectified by using feature of 'remaining masses' of both bodies towards their centers of gravity. Although these bodies have matter contents, 'remaining masses' of both bodies towards their centers of gravity reduce as the distance between centers of gravity approach zero magnitude. In reality, every basic 3D matter particle of one macro body is (apparently) gravitationally attracted towards every basic 3D matter particles of the other macro body. Due to symmetry of distribution of matter particles in both bodies, although there is positive apparent gravitational attraction between them, resultant (apparent) gravitational attraction between the bodies will reduce as their centers of gravity approach each other.

Further, we use a constant of proportion in the equation. Since there is no method to determine its magnitude, mathematically, magnitude of this constant is determined by practically measuring (apparent) gravitational attraction between many pairs of bodies of predetermined parameters. Hence, the purpose of this constant is to balance the result of the equation with the observed results, without any logical reason. Thus, the equation is valid only in conditions, where its results do not contradict rationality.

Wherever this equation may produce irrational results, the equation is considered invalid (for different reasons). However, there are certain occasions, when the inverse square law cannot provide rational results. Anomalies, noticed during estimation of (apparent) gravitational attraction, during eclipses, etc. shows that the magnitude of the constant of proportion, used in the equation, is not an absolute value.

6.1.2. Breakdown of inverse square law:

Irrespective of steady matter content of a body, its mass increases as the body's linear speed increases. Mass of a body is assumed to reach infinite level as the body's linear speed reaches the speed of light. This phenomenon takes place only in the direction of body's linear motion. In other directions, body's mass is not affected by body's linear motion. Hence, the use of mass as a parameter in the equation for apparent gravitational attraction, where one or both bodies are moving, becomes ambiguous.

If light is considered as made up of corpuscles, they are already moving at the speed of light, their masses should be infinity. Under such conditions, (apparent) gravitational attraction between them will also be of infinite proportions. All light corpuscles in the universe are bound to coagulate together under (apparent) gravitational attraction and complete darkness will be the result. This possibility is avoided by assigning mass-less stature to the corpuscles of light. Yet, bending of light beams near stars is attributed to the (apparent) gravitational attraction between corpuscles of light and the stars. Under this confused state, use of masses of high speed-bodies as parameters in the inverse square equation to find the magnitude of apparent gravitational attraction becomes useless.

As the distance between two matter bodies approaches zero, anomaly develops. At very close proximity between two macro bodies, the equation (5/7) collapses. When the distance between two bodies approaches zero, the factor 'd' in the equation also approaches zero. Mathematically, any quantity divided by zero is indeterminate or infinity. Since, practically, gravitational attraction between two bodies cannot approach infinite proportions the equation breaks down. Presently, this difficulty, in cases of general inverse square laws, is prevented by the use of 'compensating factor'. Distance between their centers of gravity is used for the distance, d, between the bodies. Since, the matter content of a body is spread out over a large volume about its centre of gravity, this is likely to produce anomalies in certain cases of very large bodies.

All interactions between 3D macro bodies are due to (the matter fields about them by) their constituent photons. Normally, matter bodies of photons in two interacting bodies do not come very close to each other. Much larger inertial pockets of the photons, enclosing their matter contents, meet long before their matter bodies can make contact. Interactions between the inertial pockets keep the matter bodies of the photons from coming in contact with each other. Consequently, distance between photons in two colliding bodies cannot approach zero. For their independent and free existence, there has to be a distance (however small it may be) between them. This means that there has to be an extent of 2D energy fields between two matter bodies. If they come close enough to get rid of this extent of 2D energy fields between them, their matter particles - two high-speed spinning bodies - interfere and destroy each other.

Angles subtended by photons, (very close to each other) of the two bodies, on each other in a 2D energy field, more or less vary directly as the distance between them. Since the distance between the bodies is already a factor of the equation (5/7), in normal cases, the variation of angle subtended by the photons of one body to the photons of other body need not be considered separately. It is accounted for by the gravitational constant. Radial sizes of all photons are the same. When the distance between two photons is comparatively large, a change in the distance between them do not make appreciable variation in the angle subtended by the photons on each other.

If the distance between two photons becomes very small, angle subtended by the photons on each other varies appreciably, when

Chapter 6. **MAGNITUDE OF ATTRACTION**

distance between them is varied. As this distance approaches zero, angle subtended on each other varies at increasingly greater rate than the rate of reduction in the distance between them. Equation (5/7) does not cater for this change, at close distances between two photons. Hence, at very small distances between two photons (matter bodies) the equation (5/7) for the apparent gravitational attraction between them breaks down, considering that the distance between the bodies is between their surfaces.

Breakdown of equation (5/7) may also be due to the following reason. Every macro body has a matter field. Matter field of a macro body is the distorted part of the 2D energy fields within the boundaries of the macro body. Because of the latticework structure of the 2D energy fields, the matter field of a macro body extends outside the limits of the physical-body for very short distance. Therefore, when two macro bodies come very close, their matter fields meet first. Every plane, passing through both the bodies, has a 2D energy field, which encompasses parts of both matter fields. As the macro bodies get nearer, each of the distorted parts in the 2D energy fields come nearer. Distortions in them, catering to different macro bodies and their states, tend to be shared. The 2D energy field squares in between the macro bodies attain an average magnitude of distortions. Total distortion is shared by all the 2D energy field latticework squares in between the macro bodies, each one in its own plane.

As the distance between two macro bodies reduces further, extent of 2D energy fields between them diminish. Number of 2D energy field-latticework squares between the macro bodies, available to share the total distortion in the matter field in that plane becomes fewer. As a result, as the macro bodies approach each other, each latticework square has to accommodate more and more distortions. Effort, required to introduce the same amount of distortion into a 2D energy field latticework square, varies in proportion to the distortion already present in the latticework square. As two macro bodies come nearer, more and more effort is required to introduce the same amount of distortions in the matter field in between these bodies. This can be taken as an increase in the resistance offered by the matter fields to accommodate the distortions. Resistance offered by the small extent of matter field in between the macro bodies reduces the effectiveness of the actions, which are moving the macro bodies towards each other. To produce the same magnitude of movement, the external efforts on the macro bodies have to be much higher, at shorter distance between the macro bodies. This phenomenon is applicable to very small macro bodies constituted by few photons.

Excessive resistance, from the 2D energy fields, hinders the motion of the bodies towards each other. Efforts, moving the bodies towards each other, become (correspondingly) less effective. In other words, the effective apparent gravitational attraction between two bodies at very close range is relatively much less than when the bodies are farther apart (over and above the effect of distance between them by the relation $1/d^2$, in the equation). This part of interaction was not taken into account, while deriving the equation for the magnitude of apparent gravitational attraction between two bodies, which are relatively large and with large distance between them.

Hence, at very close distances between two bodies, equation (5/7) for the magnitude of apparent gravitational attraction is not valid. This phenomenon prevents disintegration of basic 3D matter particles, when two 3D macro bodies collide. Reduction in the effectiveness of efforts, pushing two small 3D macro bodies towards each other, prevents their fundamental particles from coming very close. They can come in contact only in exceptional cases, when the efforts are extremely high or the 2D energy fields between the bodies collapse. In such cases, the colliding bodies interfere with each other to disintegrate fundamental particles of both the bodies. Normally, atoms or other fundamental particles of colliding bodies never come very close to each other.

Another occasion for the break down of the equation (5/7) is when there is a large difference in the rest masses of the concerned photons (very large difference in their temperatures), while considering the apparent gravitational attraction between them. Photons, making up the 3D bodies, are mostly of compatible matter content level (temperature). Their rest masses may vary very little within a small range, depending on the physical state of a macro body and state of its surroundings during stabilization periods.

Hence, taking the apparent gravitational attraction in direct relation to the number of photons in each body – rest mass of the bodies - is all right for larger 3D macro-bodies. However, when the apparent gravitational attraction, between two photons with large difference in their matter content levels is considered, it is not in direct relation to their matter contents – rest masses.

Each photon, whatever its matter content may be, takes the same time to spin through any one 2D energy field plane. Therefore, if two photons are of different rest masses, time duration for their disc planes to be in coincidence in the same plane depends only on the rest mass of the smaller photon. Hence, in this condition, equation (5/7) has to be modified by substituting rest mass of larger photon by the rest mass of smaller photon. Because of the difference in their spin speeds, it is rare for them to be in the same plane for more than an instant. This condition is applicable only when the apparent gravitational attraction between two independent photons is considered.

6.2. Anomalies in gravitational attraction:

All 3D macro bodies are transparent to most of the 2D energy fields passing through them. Transparency depends on the concentration of photons within a macro body. It is possible for some of the very large bodies to have extremely high matter density in straight lines passing through them and thereby have very high concentrations of photons, intercepting planes containing that line. An external photon, whose disc plane contains this straight line may find its disc plane intercepted not only by the disc planes of few photons in the macro body but also due to the core bodies of great many photons (in the macro body). Median plane of external photon may not coincide with median planes of these photons, which are situated in a straight line passing through its disc plane. In this case, over and above normal gravitational apparent attraction, the external photon is under additional translational action to move it towards the large body. This happens due to those photons, which are situated on the straight line but do not appear in the disc plane of the external photon. Total apparent gravitational attraction, felt by the external photon, is greater than that is proportional to the rest masses of the photon and the macro body. However, the apparent gravitational attraction felt by the large macro body is proportional to the rest masses of the macro body and the photon. Thus, a small macro body near an extremely large and dense macro body may experience greater apparent gravitational attraction, compared to the apparent gravitational attraction experienced by the large body. Total magnitude of apparent gravitational attraction between the macro body and the external photon is enhanced.

Matter density of a large macro body may vary from place to place in its body. This can make differences in the apparent gravitational attraction between this macro body and any other small body placed on or near its surface, at different places. If the smaller body is moved inside the larger macro body, magnitude and direction of resultant gravitational apparent attraction between the bodies depend on the distribution of matter in the larger macro body around the smaller body.

As shown in figure 6A, let there be three bodies A, B and D. We shall consider the apparent gravitational attraction on body 'B' towards the body 'A'. When the body 'B' is in position B_1, apparent gravitational attraction on body 'B' is along the line O_2O, connecting the centers of both the bodies. Extent of 2D energy fields, applying gravitation on body B_1 from the direction of the body 'A' are limited up to the body 'A'. On the other side, the 2D energy fields extend infinitely. Translational action on body B_1 towards the body 'A' is produced by the sum of apparent gravitational attraction between body B_1 & body 'A' (and the partial

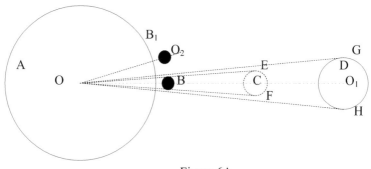

Figure 6A

gravitational attraction on photons of body B_1, whose disc planes are intercepted by flat faces of photons of body 'A').

Consider the system, when the body 'B' is in between the bodies 'A and D'. Translational action on body 'B' from the direction of body 'A' does not change. Extent of few 2D energy fields from the body 'B' towards the body 'D' is limited by the body 'D'. (Since the body 'D' is also transparent to 2D energy fields, only few 2D energy fields are blocked by its photons). Gravitational action on body 'A' from the direction of body 'D' is comparatively, reduced. There are many photons in body 'D', which may come in the median planes of the photons in body 'B'. These photons cause apparent attraction between bodies 'B and D'. Gravitational action on body 'D' towards body 'A' is correspondingly reduced by reduction in the extent of 2D energy fields and by an apparent attraction in the opposite direction. Hence, the apparent gravitational attraction between bodies 'A and D', when the intervening body is at position 'B', differs from similar attraction between them, when the intervening body is at position 'B_1'. Apparent attraction due to gravitation between bodies 'A' and 'D' is greater when body 'B' is at position 'B_1' than when it is at position 'B'.

Let another body C be placed in between bodies 'B and D'. Body 'C' will further reduce the extent of few 2D energy fields in the direction from body 'B' to body 'D'. It also produces additional apparent gravitational attraction between itself and body 'B'. These two actions together reduce the apparent gravitational attraction between bodies 'A and B'. This phenomenon produces perturbations in the measurement of apparent gravitational attraction on the surface of earth during eclipses.

Apparent gravitational attraction on a macro body is realized by its inertial action on a body. This inertial action is often understood as a linear motion of either or both of the participating macro bodies towards each other. Inertial action is produced by additional distortions introduced into the bodies' matter fields. If the distortions, introduced into macro bodies' matter field by gravitational actions, are uneven about the centre of mass of a body, whole of the inertial action will not be linear or towards the other body. A component of additional distortions, acting away from the centre of mass produce a torque on the body. In this case, component of gravitational action, available for macro body's linear motion, is less than the total work-done by the gravitational action. Hence, there will be a discrepancy between theoretical magnitude of apparent gravitational attraction between the macro bodies and the magnitude of apparent gravitational attraction available to produce body's linear motion. This phenomenon is described in detail in section 7.3, on central force.

6.3. Mass and weight:

Matter gives a body, its substance. Substance provides a body with its real existence in space. We, the rational beings, live in three-dimensional space and consider only those bodies, which have three-dimensional existence, as real. Matter in its two-dimensional and single–dimensional states are neither recognized nor treated as real. All 3D matter bodies contain matter in the form of basic 3D matter particles – the photons.

Since we have no reference to which matter content of a body can be compared with, we have no measuring system to determine matter content of a body, directly. We are compelled to rely on references derived from other measuring systems to determine matter content-equivalents of 3D matter bodies, in terms of other defined dimensional measurements. Thus, we have come to use the functional entities of 'mass' and 'weight' to represent matter content of 3D matter bodies. Both mass and weight are mathematical relations of certain parameters associated with a 3D matter body, rather than measures of matter content of a body. Hence, they are likely to show discrepancies, when body-parameters vary.

6.3.1. Mass of a body:

Acceleration, 'a', of a body in the direction of an external effort on it and the magnitude of the external force 'F' acting on the body have a mathematical relation, $(a = F \div m)$. Force is the rate of change of additional work-done about a body and

Chapter 6. MAGNITUDE OF ATTRACTION

acceleration is the rate of change of body's state of motion (velocity). We use this relation to define the functional entity, 'mass', which is generally used by physicists to represent the matter content of a body. Mass is represented by 'm' in the equation. Since the mass is only a relation, it is a functional entity with its existence in the mind of rational beings, whereas matter content of a body is a real entity with positive existence in space. Mass exists only in mathematical equations but matter content has substance and it exists in space. We use this relation, mass, to represent the matter content of a body.

At low linear speeds (with respect to an absolute reference) of a body (in normal conditions) and by assuming unity as a constant of proportion, this relation can be taken to be an (approximate) measure of body's matter content. Matter content of a larger body at rest is the sum of matter contents of all the photons in its matter field. Theoretical relationship between external effort, acting on a body at rest, and the body's acceleration is the 'rest mass' of the body. Since the body is at rest, we can say that, the 'rest mass' represents the real entity of matter but the 'relativistic mass', which is a modified adaptation according to the state of a body's linear motion is a functional entity. Only at very high linear speed of a body, difference between them becomes apparent.

Since both the force and acceleration are directional, mass of a body determined by this relation is also directional. Mass of a body is determined in the directions of external force and body's acceleration (in the direction of the force). In all other directions, mass of the body may be different. It should also be noted that in order to determine mass of a body by the above relationship, the body has to be accelerated by the external effort. If full or part of an external effort acts on the body in any other direction, it is ignored. Mass of the body, determined thus, may have little relationship to body's real matter content.

As the linear speed of a body (its matter field) approaches the speed of light, relation between force, mass and acceleration, as given by the above equation, breaks down. Matter field, of a moving body, contains additional distortions, required to produce its linear motion at a constant linear speed. Because of the stress, already present in the matter field-latticework squares, effort required to produce more additional distortions in them is proportionately higher. Distortions in the matter field latticework squares, leaving the matter field at the rear end of the moving body, have reactions and the rearward speed of the distortions is proportional to the additional distortions contained in the matter field. An external effort, in the direction of motion of the body, need to overcome this reaction and should be able to produce a forward motion of quanta of matter in the latticework squares, which are expanding in rearward direction. If the matter field of a body contains larger magnitude of additional distortions, it will need larger external effort to overcome this reaction and to introduce more additional distortion into the matter field of the body, to increase body's linear speed. That is to say, to make the same speed variation (acceleration) of a body, the external effort (force) needed is greater at higher linear speed of the body.

This is not in accordance with the relation, ($a=F/m$), between them, as given above, while defining the rest mass of a body. If we take the above relation between the external force and the mass to be infallible, difference in this relation at higher speed, produces an appearance that the mass of the body has increased at higher speed, without a corresponding increase in its matter content. Since we have no method to determine actual matter content of the body, we presume the difference in the relation between external effort and acceleration is due an increase in its matter content. This apparent increase in the matter content of a body is the 'relativistic mass' of the moving body. This is effective only in the direction of motion of a body. There is no change in the matter content of the body (unless speed of the body exceeds the speed of light) and hence, the rest mass of the body remains constant.

In fact, this apparent increase in the mass of a body is not due to a change in the body's matter content at all. It is produced due to the inability of the external effort (force) to affect the linear motion of an already moving body; at the same rate, as it affected the linear motion of the same body, when it was at rest or moving at very slow linear speed. It is an error introduced by our assumption that the acceleration of a body is directly proportional to an external effort applied on it. As is seen here, their relationship is not direct and strictly speaking, this relationship does not represent the matter content of the body, other than at low speeds of the body. This phenomenon is produced due to the latticework structure of the 2D energy fields. As the speed of a body increases, inefficiency of the external effort (force) on the body in its direction of linear motion also increases. This gives increased relativistic mass at higher linear speed of a body.

Force is the rate of work-done about a body. It is the measure of additional distortions, introduced into the matter field of a body in unit time. If the matter content of a body remains constant, a change in its mass could be introduced only by a variation in the force. Relativistic mass of a body, moving at high linear speed is a product of depreciation in the magnitude of force (rate of work done in the matter field of the body) rather than an increase in its matter content. This only means that as the linear speed of a body increases, efficiency of an external effort on the body in the direction of its linear motion depreciates.

Quanta of matter, at their junctions in a 2D energy field, are loosely held together by their mutual (apparent) attraction between their matter contents and by the push force exerted by the compressed quanta-chain. Their bond is not rigid. There is a maximum speed limit on the movements of quanta of matter in the 2D energy fields' latticework structures. Should the speed of motion of the quanta of matter exceed this limit, the 2D energy fields will (locally) breakdown. If a body attains speed of light, an external effort (however large it may be) in the direction of motion of the body cannot produce additional distortions in its matter field. At this stage, it may be presumed that, in the direction of force, mass of the body has become infinity. In reality, nothing has happened to the body. It is that no external effort in the same direction can accelerate the body further. Because of this, at present, it is erroneously assumed that all the effort (energy) applied to the body is converted to mass (matter) and body's mass has become infinity. Instead, it is the action of the effort on the body, which is the force that has reduced to nil. Though the external effort is applied to the body, corresponding force does not develop on the body any more. If the direction of the external effort is different

Chapter 6. MAGNITUDE OF ATTRACTION

from the direction of the motion of the body, only its component in the direction of motion of the body becomes ineffective to produce relativistic mass.

A macro body is constituted by numerous photons. Each photon moves at its critical speed within the body with respect to surrounding matter field. When the macro body is moving in any direction, its matter field moves at the same speed and hence these photons' critical speeds are not affected by the linear speed of the body. There are millions of photons in a macro body. They move at their critical constant linear speeds within the body's matter field, in circular paths in various directions in different planes. An external effort, acting on the macro body tends to accelerate macro body's matter field. It is not acting to increase photons' linear speed. Hence, this phenomenon of relativistic mass does not affect constituent photons of a macro body, which are already moving at the speed of light.

Mass of a body is also understood as the quantitative measure of inertia of a body to changes in its state of linear motion. Inertia represents the resistance offered by a body to a change in its state of linear motion. Another functional entity, 'moment of inertia', is used to represent the resistance offered by a body to a change in its state of rotational motion. Moment of inertia is a quantitative measure of rotational inertia. Moment of inertia of a body is with reference to an axis within or outside the body material about which the matter body is rotating. Hence, its magnitude depends on the reference chosen. It is the measure of resistance offered by the body to a torque acting on the body to rotate it. Moment of inertia is defined as the sum of the products obtained by multiplying the mass of each particle of matter in a given body by the square of its distance from the axis. However, unlike mass, moment of inertia of a body has comparatively very little relation to the body's matter content.

Mass of a body is generally used to represent its total matter content. Thus, mass has become synonymous with matter content of a body. Average location-point of total matter content within a macro body is considered as its 'centre of mass'. In mathematical treatments, this imaginary point may be assumed as the seat of

Chapter 6. MAGNITUDE OF ATTRACTION

whole matter content of the macro body in concentrated form. Centre of mass can also be used to represent the location of total matter content of a system of independent macro bodies.

However, if the term 'mass' is understood in its literal sense, it is the relation between the magnitude of work contained about a body and the body's velocity. If the velocity of a body cannot be increased by an external effort, it only means that no work is done about the body, irrespective of the presence of external effort.

It is physically impossible for an external effort to act through a point. Usually, an external effort acts through a wider area and often through the whole area of the macro body. If the external effort, acting on a macro body, is uneven about its 'centre of matter content', matter particles on either sides of the plane through the centre of matter content experience different magnitudes of acceleration in the direction of effort. In this case, the centre of mass of the macro body shifts towards the side of matter particles with higher acceleration.

6.3.2. Weight of a body:

In order to determine matter content of a small 3D macro body, in the proximity of (much larger 3D macro body's) earth's surface, a functional entity termed as 'weight' is used. Weight is the magnitude of apparent gravitational attraction between the small macro body and the earth. Normally, the factors affecting the weight of a small macro body (earth's total matter content and the distance between centre of gravity of body and centre of gravity of earth) are considered constant. Knowing the acceleration of a small macro body under free fall, we are able to determine the magnitude of (apparent) attraction due to gravitation - weight - between the small macro body and the earth by using the equation, ($F=ma$). This value is further converted to gravitational units by dividing the right hand side-factors of the equation by a predetermined value of acceleration due to gravitation near earth's surface, 'g'. In this case; ($g=a$). Thus, the weight of a body is able to give us the numerical equivalent of the rest mass of the body and that is generally taken as equivalent to the matter content of the small macro body.

Chapter 6. MAGNITUDE OF ATTRACTION

Weight of a body may also be understood to be the effort required to support a body in a relatively static condition with respect to the surface of the earth, from moving towards earth's center of gravity. Full weight of the small macro body can be obtained only when the theoretical acceleration of the body due to apparent gravitational attraction between the body and the earth is fully neutralized by a theoretical acceleration provided by the reaction from the support or a restricting effort on the small macro body's fall towards the center of gravity of earth. Magnitude of this external effort, when converted into gravitational unit, appropriate to earth's surface, gives us the measure of weight of the body.

Consider a small macro body, accelerating towards a large macro body under the action of mutual apparent gravitational attraction. (For the sake of this discussion, we shall ignore acceleration of the larger macro body and consider that the acceleration of the smaller macro body is the combined action of the accelerations of both the macro bodies). Smaller macro body continues to be under acceleration due to gravitation until it merges with (the centre of) the larger macro body. When the smaller macro body is free to accelerate towards (the centre of gravity of) the larger macro body, it is assumed to be under free fall. Since the small macro body is not restricted (supported) by an external effort, it appears to be 'weightless'. If the supporting effort, against the apparent gravitational attraction, is more than that is required to prevent small macro body's acceleration towards (the centre of) the larger macro body, the weight of the small macro body will be proportionately higher. This is how a person in an accelerating rocket feels higher gravitational action towards earth (person's weight).

Action of an external effort on a macro body, in the direction of its linear motion and the body's acceleration also depend on the present linear speed of the body. Therefore, as the velocity of a small macro body towards a larger macro body increases, effect of apparent gravitational attraction on the small macro body decreases. Magnitude of its acceleration declines. However, the body continues to increase its velocity at a slower rate. This process will continue until the velocity of the small macro body reaches a stage when its body-particles breakdown to primary particles. Thus, many of the smaller bodies, accelerating in space towards a larger body, normally revert to their constituent primary particles long before they attain the velocity of light. Liberated primary particles of the disintegrated body move away in various directions, depending on the direction of their motion at the instant of liberation. This phenomenon reduces the probability of too many small bodies from the outer space, bombarding earth or any other larger bodies in space. Many of the smaller celestial bodies, which are able to attain linear speeds, nearer to the speed of light, disintegrate before they can approach the earth.

Should there be an additional external effort (force) on a smaller falling body in the direction of its motion; the body will have a supplementary acceleration due to the additional external effort. This is subject to the condition that the force-applying body/mechanism is moving faster than the body itself. If the body is free, it will be already moving in free fall and be weightless. Additional external effort on the body is trying to accelerate the weightless body in the direction of its motion. While doing so, a reaction is developed in the body's matter field. This reaction acts as the weight of the body in the negative (opposite) direction. That is, now the body has a weight in the opposite direction − away from the larger body. This is the reason why the weight of a high speed spinning body (a gyroscope whose axis is not vertical) varies. One half of the spinning body that is moving toward the earth produces a weight reduction and the other half of the spinning body, moving away from the earth produces a weight gain, at any instant. Magnitude of weight-variation depends on the linear speeds of points on the body. Magnitude of weight-variation will (theoretically) reach maximum, when instantaneous linear speeds of the peripheral points of the body reach the speed of light.

An imaginary point in space, where the total weight of a macro body is concentrated is called the body's 'centre of gravity'. Depending on the shape and distribution of its matter content, centre of gravity of a macro body may be within or outside the perimeter of the body. [Refer section 6.1].

Weight of a body is the magnitude of restriction needed to

nullify its linear acceleration (towards earth). Similar actions taking place even outside the influence of earth can create weight of a body. Restriction to linear acceleration of a body by any means can cause production of weight about the body. Only difference is that in the former case, both the acceleration (provided by action of apparent gravitational attraction towards earth) and the restriction take place near earth. In the later case, both the acceleration (provided by an external effort) and the restriction are measured without considering the presence of earth or apparent attraction due to gravity. Hence, there is no difference between 'gravitational mass' and 'inertial mass', as differentiated today.

6.3.3. Matter contents of fundamental particles:

All 3D matter particles, constituted by very few photons; bitons, tetrons and hextons (electrons and positrons) are included as fundamental particles, in this paragraph. Constituent photons of these particles have equal matter contents. Rest mass of a fundamental particle is the total rest mass of its constituent photons.

A biton's rest mass is twice that of its photons. Tetron's rest mass is four times that of a constituent photon. A hexton's rest mass is six times that of its constituent photons. Photons of a biton spin synchronously, while moving in the same curved path. Photons' median disc plane (in a biton) change continuously. Normally only one of the bitons of the tetron/hexton has its photons in one plane at a given instant. Therefore, the rest mass of a tetron/hexton or of a biton determined by a measure of the apparent gravitational attraction between the biton/tetron/hexton and another body (converted into gravitational unit - weight) is only due to the photons of one of the bitons. This gives part-measurement of its actual rest mass. Rest mass, thus determined, is an average measure proportional to the actual rest mass. Apparent attraction due to gravitation, in a plane, lasts only for an instant and appears only twice every spin of the photons. If the measured value is compensated by using the gravitational constant, determined for macro bodies, result will be very minute compared to its actual value.

In a tetron/hexton, there is one instant during every spin of its photons, when all of its photons come in the same plane. Weight of the tetron/hexton, determined in this plane, corresponds to tetron's/hexton's actual rest mass. In all other planes, the rest mass determined by measure of weight of a tetron/hexton, is only half/one-third of this value. Thus, its is possible to get two results, in all types of very small subatomic particles, when determining their rest mass by methods relying on the apparent attraction due to gravity towards a larger macro body. These values depend on the relative direction of the planes of the biton/tetron/hexton and the reference plane. Even when other methods are used, effect of the apparent attraction due to gravity, in this manner, is liable to show up.

Other methods to determine rest mass of a tetron/hexton depend on the interaction between one of its distortion fields and another external distortion field. Distortion fields of a tetron are highly directional. Even in a hexton, its distortion fields are directional in close proximity of the body. Most effects of the distortion fields are concentrated in few parallel planes coinciding with the planes of its bitons. These distortion fields are in different directions in different quadrants of a tetron. All other planes have only a minor share of partial distortion fields. Hence, the rest mass of a tetron/hexton (or that of very small 3D matter bodies), determined in relation to its distortion fields is not reliable. In larger bodies, such differences need not be separately considered, due to large margin of error allowed and prolific number of tetrons/hextons in the body.

* ** *** ** *

Chapter SEVEN
GENERAL

7.1. Aether drag:

It was the failure to notice an assumed aether drag on earth's motion through space that ended the progress in the search for an all-encompassing medium. This was unnecessary because the assumption of 'aether drag' itself is unwarranted. In the explanations above, it was shown that every basic 3D matter particle is moved by 2D energy field at the highest possible speed. Basic 3D matter particles (constituting primary particles, fundamental particles and higher matter bodies) move in curved paths, within a composite body. Sizes of their circular paths are limited within primary particles of the macro body. A macro body consists of millions of basic 3D matter particles – photons – within its body. Movement of a macro body is achieved by simple displacements or deflections of photons' circular paths in space, without affecting their inherent motions. 2D energy fields produce displacements or deflections of photons' circular paths. Matter has no ability to move on its own. Since the '2D energy fields' is the mover (which is displacing macro body), there will be no relative motion or friction between them. Action is limited to 2D energy fields within and in the immediate neighborhood of the macro body.

Motion of a matter body through the 2D energy fields is like the motion of a body, floating in a narrow ocean current. Ocean current carries the body along with it and there is no relative motion between the floating body and the surrounding water in the current. However, this floating body has a clear relative motion with respect to the vast ocean, outside the current. Similarly, it is the moving distortions in the 2D energy fields, which is moving a matter body. This part of 2D energy fields is a local region in and about the matter body. Distortions carry the matter body along with it and there is no relative motion or friction between them. However, with respect to the vast 2D energy field, outside body's matter field, the matter body has a relative displacement in space. 2D energy fields do not restrict but move the matter bodies through them.

Photon's linear speed is the highest possible linear speed through 2D energy fields that can be sustained without breakdown of 2D energy fields. Photons of primary particles, during their motion in curved paths through the 2D energy fields, experience resistance (opposition to their linear motion in curved paths) from the 2D energy fields. Photons' ejection (moving force) is also caused by the 2D energy fields. Since the resistance from 2D energy fields is already accounted for in the inherent motion of the photons, such resistance will not be carried further to affect the motion of 3D macro bodies. Therefore, 3D macro bodies will not experience additional drag to their motion through 2D energy fields (space). This makes the assumption of aether-drag unwarranted.

7.2. Heat and gravitation:

Heating or cooling a macro body changes its matter content and size. Energy associated with a macro body also changes corresponding to changes in its matter content. Contrary to present belief, matter content and energy content of a macro body is highest, when the macro body is coolest and in free space. Conversely, matter content and energy content of a macro body reduce, when the body is heated or when it is in the neighborhood of other 3D macro bodies. [See parent book for more details].

Changes in the matter content of a macro body affect its response to an external effort or to the apparent gravitational attraction differently. An external effort is applied into the matter field of the macro body, while the apparent gravitational attraction is the resultant of 2D energy fields' direct actions on the matter cores of constituent basic 3D matter particles. Although both are external efforts, mode of their application causes a difference in the magnitude of macro body's response.

7.2.1. Heating:

We shall consider a macro body in gaseous state for illustrations. When a gaseous macro body is compressed, external pressure reduces its volume. Smaller constituents of the macro body are brought nearer and they are held in that relative position against natural efforts (forces), trying to take these constituents back to their regular distances (between them) in their natural formations. This pressure energy, invested in the macro body to reduce body's volume, is held within the body's matter field until compression of macro body is removed and the body attains its original volume. During the reduction of its volume, a macro body under compression is heated without any other external influences as can be noted by increase in its temperature. Body radiates matter in the form of photons (infrared or higher frequency radiations). [In currently prevailing theoretical terms, during the heating of a macro body it loses energy in the form of heat radiation.] However, according to the concept in the parent book, on heating a macro body, it loses both matter and energy contents from the body. Matter content and energy content lost from the macro body are approximately proportional to each other. Temperature of a macro body is its matter content level with respect to the matter content level of a reference body.

Gradually the macro body loses enough matter content through radiation that its temperature returns to room temperature. Energy input or work-done in association with the macro body, to compress it, has not changed. Nevertheless, the macro body (by heating) has lost some matter content and energy content proportional to the matter content lost. Matter content, lost from the macro body, is not originated or converted from pressure energy, applied on body's matter field. Matter content, lost from the body, is not related to the pressure energy put into the body's matter field. Thus, heating a macro body reduces both its matter and energy levels and thereby reduces its total matter and energy contents. This is contrary to common belief that during heating, energy level of a body increases.

Conversely, when external pressure on a macro body is reduced, the macro body cools down. That is, it takes-in matter content from surrounding 2D energy fields. Corresponding energy content is developed in the body's matter field. External pressure on a macro body is least when the body is in free space. Because, in free space, there is no other macro body in the vicinity to influence the 2D energy fields surrounding the macro body. In free space, a macro body will be coolest and at its highest matter and energy content levels. Cooling a macro body increases its matter and energy content levels and thereby increases its total matter and energy contents. Temperature of a macro body is generally taken as an indication of its matter content level and energy content level. Contrary to present belief, higher temperature indicates lower energy level and lower temperature indicates higher energy level of a macro body. [Detailed mechanism of heating and cooling is given in the parent book].

7.2.2. Acceleration due to external force:

Changes in the matter contents of primary particles, in a macro body, due to the difference in its temperature, affect inertial actions of a macro body under external efforts. Magnitude of external force divided by the magnitude of acceleration is the rest mass of a (static) macro body. Rest mass of a macro body is assumed to represent its matter content. In these calculations, variations in the matter content of the body, under changes of temperature, are not considered. Since a change in the temperature of a body changes its matter content level, rest mass of the macro body also changes.

Let an external effort on a macro body is of constant magnitude. Let this effort accelerate the macro body, whose temperature varies. At higher temperature, the macro body has less matter content and hence the acceleration of the body will be of greater magnitude. This corresponds to a reduction in macro body's rest mass. Similarly, at lower temperature, the macro body has higher matter content level. Its acceleration under the same external effort will be of lower magnitude. This corresponds to an increase in the macro body's rest mass. Thus, under the action of constant external force, a macro body at higher temperature will have higher acceleration compared to the acceleration of the same body at lower temperature, under the action of identical external effort.

7.2.3. Acceleration due to gravity:

Consider a small macro body near a large macro body.

Apparent gravitational attraction between the macro bodies takes place, whenever the disc planes of constituent photons of both the macro bodies coincide. Abundant photons in both macro bodies maintain an average magnitude of apparent gravitational attraction between them. Changes in the matter content of a photon changes the angular thickness of its core-disc segments. Higher matter content increases and lower matter content reduces the angular thickness of photons' core-disc segments. As the smaller macro body is cooled, its matter content level increases. Angular thickness of core-disc segments of its photons and the frequency of their spin increase corresponding to the increase in the matter content level. These changes increase the angular sweep area of the photon-segments and increase the number of instants of apparent gravitational attraction between the macro bodies, due to each of its photons. As the smaller macro body is cooled its apparent gravitational attraction, towards the larger macro body, increases. Opposite conditions occur, when the smaller macro body is heated.

Let the rest mass of small macro body is 'm' and rest mass of the large macro body is 'M'. 'G' is the gravitational constant in 3D space system and 'd' is the distance between the centers of gravity of the macro bodies. We shall use equations, used in 3D spatial system, for the following explanations. We may neglect action of apparent gravitational attraction on the larger macro body.

Magnitude of apparent gravitational attraction, GF, between the macro bodies at a reference temperature,

$$GF = MmG / d^2$$

GF is the accelerating force on the small macro body.

Accelerating force = Rest mass × acceleration

Substituting apparent gravitational attraction in this equation;

$$MmG / d^2 = ma \qquad (7/1)$$

Where 'a' is the acceleration due to gravity of small macro body towards larger macro body.

$$a = M G \div d^2 \qquad (7/2)$$

Let the increase in rest mass due to enhancement of sweep area of the photon-segments, during reduction in temperature, is proportional to $(K_1 t)$. Mass of the small macro body increases to '$m(K_1 t)$', where K_1 is the constant of proportion and 't' is the change in temperature.

This increment in rest mass affects both sides of the above equation equally. Let the distance between the centers of the bodies remain constant.

Apparent gravitational attraction, $GF = Mm(K_1 t)G \div d^2$

Putting these values in equation (7/1);

$$Mm(K_1 t)G \div d^2 = m(K_1 t) \times a$$

$$a = MG \div d^2 \qquad (7/3)$$

Equation (7/3) is the same as equation (7/2). Hence, increment in the rest mass of the small macro body, due to reduction in temperature (or due to any other phenomenon) does not affect its acceleration due to gravity towards the larger macro body. However, their increased matter content makes the photons in the smaller macro body to spin faster. Due to increase in their spin speeds, disc plane of each photon in smaller macro body coincides more frequently (in unit time) with the disc planes of photons in the larger macro body. This increases the average magnitude of apparent gravitational attraction between the macro bodies.

Let the increase in apparent gravitational attraction due to enhancement of photons' frequency is proportional to $K_2 t$, where K_2 is constant of proportion and 't' is the change in temperature.

Magnitude of apparent gravitational attraction, GF, between the macro bodies;

$$GF = Mm(K_1 t)(K_2 t)G \div d^2 \qquad (7/4)$$

Putting value of rest mass of smaller macro body as '$m(K_1 t)$' and the value of external force, GF, from equation (7/4) in equation (7/1);

$$GF = Mm\,(K_1t)(K_2t)G \div d^2$$

$$Mm\,(K_1t)(K_2t)G \div d^2 = m(K_1t) \times a$$

$$a = M\,G(K_2t) \div d^2 \qquad (7/5)$$

In this case, the magnitude of acceleration due to gravity is higher by a factor (K_2t), compared to equation (7/3). Acceleration due to gravity of a smaller macro body towards a larger macro body increases as the smaller macro body is cooled. Reverse action takes place, when the temperature of the smaller macro body is raised. A (small) hot macro body has lesser acceleration due to gravity towards a larger macro body compared to the same body in cooler state.

Acceleration due to gravity, in gravitational unit, is the weight of a macro body. Gravitational unit, determined for a large macro body is assumed constant. Hence, an increase in acceleration due to gravity of a smaller macro body towards the larger macro body effectively increases small macro body's weight. Thus, weight of a small macro body, near a large macro body, increases as the small body's temperature is lowered (the body is cooled). Conversely, a reduction in acceleration due to gravity of a smaller macro body towards a larger macro body effectively reduces small macro body's weight. Thus, weight of a small macro body, near a large macro body, decreases as the body's temperature is raised (the body is heated). Changes in the weight of average sized macro bodies may be too small to be observed in normal conditions.

7.3. Central force:

In its stable state, a large group of free macro bodies, constituting the galactic system, as a whole may remain static in space. All free macro bodies in a galaxy are always in linear motion. Even if they are moving in curved paths, their linear motion is maintained by inertial properties of their matter fields. They are usually far out of each other's matter field. They do not influence each other, other than for their mutual apparent gravitational attraction. Mutual apparent gravitational attraction between such bodies influences their paths of linear motion.

Two macro bodies of suitable parameters, moving in space, influence each other's linear paths under the action of mutual apparent gravitational attraction. These bodies will appear to move in association with each other and form a system of multiple bodies. They are said to be orbiting about each other. Two macro bodies, orbiting about each other is a binary system. A group of more than two bodies, orbiting about each other is a planetary system. If one of the bodies of a group is imagined to be stationary in space, the other bodies of the group appears to orbit around the apparently stationary body. The body that is considered static in space is the central body and the other bodies in the group are the planetary bodies. A planetary body appears to orbit around its central body in circular or elliptical (forming a path of closed geometrical figure) orbital path. Influence of apparent gravitational attraction between a planetary body and its central body appears to be towards the central body, i.e., towards the centre of its circular path. Hence, in this case, the apparent gravitational attraction between a planetary body and its central body is called a 'central force'.

In this section, all actions of a planetary body due to its inherent inertial motion are credited to its linear motion/work and all actions due to the central force are credited to radial motion/work. A body is defined by measurements of space, occupied by its matter field and by its rest mass, representing its matter content. A free body tends to move in a straight line due to inertial properties of associated matter field. Apparent gravitational attraction between two bodies is the result of apparent gravitational attraction between their basic 3D matter particles. Inertia of a body does not apply effort (a force) on the body. While inertia maintains a moving body in a straight line-path, it is the central force by its action on the body, which changes the direction of a planet's linear motion and produces its spin motion. Actions of each body are between it (its basic 3D matter particles) and the surrounding 2D energy fields. Concurrent actions on two bodies, considered together, may be interpreted as an apparent interaction between them. Although gravitational action on each body is separate, such actions on the central and planetary bodies, when considered together, provide a central force of apparent gravitational attraction between them. A planet is apparently attracted towards the central body and vice versa. Apparent gravitational attraction between two bodies, at any instant takes place only in common planes occupied by the disc-planes of their constituent photons.

Actions, similar to the orbital motion of a planetary body, considered here, takes place on central body also. All actions in space take place with respect to absolute reference. Since the bodies under consideration are far out of each other's matter field and each one is independent of the other, their motions are normally correlated to get 'relative motion' with respect to one body, which is usually considered as static in space. With respect an absolute reference (the 2D energy fields), each body has its absolute motion in space.

A moving macro body (its matter field) contains additional work, required for its linear and spin motions (disregarding all work required to sustain body's integrity and any local phenomena, it may have). A free macro body, which is associated with such work, continues its linear motion in a straight line and its angular (spin) motion at constant speeds. Additional work, contained in an orbiting macro body's matter field, was invested into its matter field by external efforts, including the apparent gravitational attraction towards the central body, before its entry into its orbital path. Another external effort is required to change the state of any of its constant motions. To keep changing the direction of its linear motion, appropriate to the orbital path, distortions in its matter field are modified continuously. Changes in the matter field distortions produce a body's accelerating stage. Instantaneous velocity of an orbiting body depends on the magnitude of additional matter field-distortions, its matter field contains at that instant. Acceleration of the orbiting body depends on the variation in the magnitude or direction of additional distortions in body's matter field. [In this paragraph, the term 'force' is used in its general sense to represent an effort or the cause of an action]. Forces in different planes do not interact. They act on the matter particles of a macro body independently, in each plane. Matter particles are moved by each of the forces in its own direction and in its own plane to produce resultant motion of the matter particles in 3D space.

Since there are no static macro bodies, in space (within a galaxy), no free orbiting body can orbit around another body (A free orbiting body is that which is free from all external influences other than the central force towards a central body). An orbiting body and the central body orbit about each other by the changes in the directions of their (real orbital) paths. A macro body, considered as the main body of a system of bodies, may be called as the central body and other macro bodies in the system may be called the planetary bodies. With respect to a central body, assumed motionless, planetary bodies appear to move around the central body in their apparent orbits. A planetary body, simultaneously, maintains a constant acceleration towards the central body, maintains its motions at constant velocities in linear and radial directions and maintains a spin motion.

A planet has two simultaneous linear motions, a linear motion nearly tangential to its orbital path and a radial motion towards the central body. Linear motion of the planet, at any point on the orbital path, is directed away from the tangent to the orbit. Angle between the direction of linear motion and the tangent at a point on the orbital path produces an outward perpendicular component of the planetary body's linear motion. This is a real motion of the body, which replaces the assumed motion produced by the 'imaginary centrifugal force' on the body. Major part of the work (linear matter field-distortions), within a planet's matter field, carries the planet along its (real) orbital path. Relative direction of the radial motion to the tangential linear motion varies at different points on the real orbital path. Most of the matter field-distortions, producing linear motion and radial motion of the body, are in different planes. Hence, they do not produce resultants. However, independent displacements of the body, produced by the matter field-distortions in different planes, together, may be understood as the resultant displacements (motion) of the planet along its orbital path.

Since there are no static bodies in nature, in the following analysis, a planetary orbit about a moving central body (moving in a much larger circular path) is considered. As a planet moves in its orbit about the central body, its relative direction to central body changes through half a circle, alternately in either side of central body's path. This cyclic shift in the relative directions about the central body's path appears as apparent orbit around the central body. This is in contrast with present assumption of a planet moving in full circles/ellipses around its central body. Changes in relative direction between central and planetary bodies cause variations in the magnitudes of forces and their actions. Explanations given below are for relative position of these bodies, at the instant, when they are moving in parallel linear paths in the same direction. The planetary body is considered, at a point in its orbit, when its linear motion is in parallel direction to the central body's direction of motion and the planetary body is over-taking the central body in the direction of its linear motion.

Central force, between a planetary body and its central body, is provided by the apparent gravitational attraction between them. Direction of this apparent attraction is, currently, considered as perpendicular to the direction of planet's linear motion. This assumption is suitable only for those planetary bodies, whose orbital path is around the central body, like a planetary body attached to the central body by an un-stretchable link. Since there is no mechanical link between a central body and the stellar planets, they do not orbit around their central bodies but they orbit about their central bodies under the action of central force. But when the observer's platform on the central body is assumed to be static in space, their orbital paths appear to be around the central body.

Action of the central force depends on the magnitude of (radial) matter field-distortions, it is able to invest into the matter field of the planetary body. Magnitude of matter field-distortions (in perpendicular direction to its linear motion) that a macro body is able to store is governed by the magnitude of the perpendicular force and the body's absolute linear speed. Gravitational action is instantaneous and continuous. As long as the participating bodies occupy common planes, apparent gravitational attraction on the bodies continues to invest additional distortions in their matter fields.

Figure 7A shows a (homogeneous) spherical planetary body of radius 'r', with its center at O_2 and moving to the left (in the figure) in relation to the central body. We shall consider the front hemisphere of the planetary body, shown in the figure by the segment APCQBO$_2$A. The planet moves in the direction from O$_2$ to C, in its linear (orbital) path. Absolute linear speed of the body being V m/sec, whole body takes 2r/V seconds to pass a point in space. Due to difference in sizes of the central and planetary bodies, it takes more than a unit time for the planet to move across the central body, in any tangential direction to the planet's orbital path. Gravitational effort acts on the planet in perpendicular direction to the orbit, parallel to the line AB, for whole of this time. Work introduced by the central force in a cross section of the planet depends on the time during which it is under gravitational influence in a particular direction. Density of (perpendicular) matter field distortions increases from C to O$_2$.

Take an elementary circular section PQ (cut by planes parallel to AB at distances x and x+dx from AB) of thickness dx, perpendicular to the line XX. Magnitude of matter field distortions in any part of the body, in its steady state, is proportional to its volume.

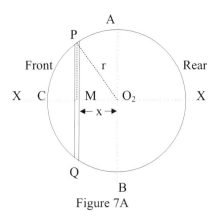

Figure 7A

$$(PM)^2 = (O_2P)^2 - (O_2M)^2 = r^2 - x^2, \qquad CM = r - x,$$

Volume of section PQ = π ($r^2 - x^2$) dx

Matter density of section PQ = $m \div \dfrac{4\pi \times r^3}{3} = \dfrac{3m}{4\pi \times r^3}$ \qquad (7/5)

Matter content of section PQ $= \dfrac{3m}{4\pi \times r^3} \times \pi(r^2 - x^2)dx$

$= \dfrac{3m(r^2 - x^2)dx}{4r^3}$

Using inverse square law for apparent attraction due to gravitation;

Central force on section PQ $= \dfrac{3m(r^2 - x^2)dx}{4r^3} \times \dfrac{MG}{D^2}$

$= \dfrac{3MGm(r^2 - x^2)dx}{4r^3 D^2}$ (7/6)

Where, M is the mass of the central body, G is the gravitational constant in 3D space system and D is the distance between the centers of gravity of section PQ and the central body. Central force is the rate of investment of additional (perpendicular) matter field-distortions into section PQ of the planetary body with respect to the rate of rate of distance moved, towards the central body.

Central force acts on section PQ for the time, during which it exists under the force. Since, we are considering the motion of the planet across its orbital path; we are interested only in those common planes (perpendicular to planet's orbital path) containing both the central and planetary bodies. As soon as the front edge of the planet reaches the perpendicular line passing through the rear edge of the central body, both bodies start to have common planes (in the direction, considered). The number of common planes increases as the planet moves forward to overtake the central body. Duration of action of the central force on the planet, in perpendicular direction to orbital path, before the section PQ is in the common planes with the central body is from the time planet's forward edge enters the common plane with the central body to the time, when the section PQ enter the common plane.

Distance between the front edge of the planet and section;

PQ = (r − x).

Time duration = displacement / speed = $(r - x) / V$

Let the constancy of proportion between force and magnitude of additional distortions introduced by it is equal to 'k'. This constant of proportion for each planetary body is different. It depends on the size of the planetary body in the direction of force, its consistency and matter density.

Magnitude of (radial) matter field-distortions invested in PQ;

$$= \frac{3MGm(r^2 - x^2)dx}{4r^3D^2} \times \frac{(r-x)}{V} \times k$$

$$= \frac{3MGmk}{4r^3D^2V}(r^2 - x^2)(r-x)dx$$

Magnitude of total (radial) matter field-distortions in the hemisphere $ACBO_2A$ of the dynamic planet, when the whole planet is within the common planes with the central body,

$$W_1 = \sum_{x=0}^{x=r} \frac{3MGmk}{4r^3D^2V}(r^2 - x^2)(r-x)dx$$

$$= \frac{3MGmk}{4r^3D^2V} \int_{x=0}^{x=r}(r^2 - x^2)(r-x)dx$$

$$= \frac{3MGmk}{4r^3D^2V} \int_{x=0}^{x=r}(r^3 - r^2x - rx^2 + x^3)dx$$

$$= \frac{3MGmk}{4r^3D^2V} \left| r^3x - \frac{r^2x^2}{2} - \frac{rx^3}{3} + \frac{x^4}{4} \right|_0^r$$

$$= \frac{3MGmk}{4r^3D^2V} \left(r^4 - \frac{r^4}{2} - \frac{r^4}{3} + \frac{r^4}{4} \right)$$

$$= \frac{3MGmk}{4r^3D^2V} \times \frac{5r^4}{12} = \frac{5MGmkr}{16D^2V}$$

(For the time being, the constant of integration is neglected). Since the value of the gravitational constant G is determined experimentally, we can take that the operation by the constant of proportion, k, is also automatically accounted for in the value of G. Hence, we may neglect the factor k in the above equation.

$$\text{Thus,} \qquad W_1 = \frac{5MGmr}{16D^2V} \qquad (7/7)$$

Similarly, taking other hemisphere AO_2BEA, of the planetary body as shown in figure 7B;

Take an elementary circular section PQ (cut by planes parallel to AB at distances x and x+dx from AB) of thickness dx, perpendicular to the axis XX. Magnitude of matter field distortions in any part of the body, in its steady state, is proportional to the volume of the part and its distance from C, forward end of the body.

$$(PM)^2 = (O_2P)^2 - (O_2M)^2 = r^2 - x^2,$$

$$CM = r + x,$$

$$\text{Volume of section PQ} = \pi(r^2 - x^2)dx$$

$$\text{Matter density of section PQ} = m \div \frac{4\pi \times r^3}{3} = \frac{3m}{4\pi \times r^3} \text{ kg/m}^3$$

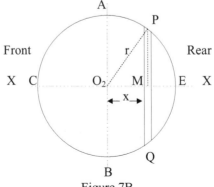

Figure 7B

Matter content of section PQ;

$$= \frac{3m}{4\pi \times r^3} \times \pi(r^2 - x^2)dx = \frac{3m(r^2 - x^2)dx}{4r^3}$$

Central force on section PQ;

$$= \frac{3m(r^2 - x^2)dx}{4r^3} \times \frac{MG}{D^2} = \frac{3MGm(r^2 - x^2)dx}{4r^3 D^2}$$

Central force is the rate of investment of matter field distortions in the section PQ, in the direction of central body.

Time duration in which PQ is under central force

$$= (r + x) \div V$$

(Taking the constant of proportion equal to k),

Magnitude of (radial) matter field distortions invested in PQ;

$$= \frac{3MGmk(r^2 - x^2)dx}{4r^3 D^2} \times \frac{(r + x)}{V}$$

$$= \frac{3MGmk}{4r^3 D^2 V}(r^2 - x^2)(r + x)dx$$

Total magnitude of (radial) matter field distortions in the hemisphere AO2BEA of the dynamic planet,

$$W_2 = \sum_{x=0}^{x=r} \frac{3MGmk}{4r^3 D^2 V}(r^2 - x^2)(r + x)dx$$

$$= \int_{x=0}^{x=r} \frac{3MGmk}{4r^3 D^2 V}(r^2 - x^2)(r + x)dx$$

$$= \frac{3MGmk}{4r^3 D^2 V} \int_{x=0}^{x=r}(r^3 + r^2 x - rx^2 - x^3)dx$$

$$= \frac{3MGmk}{4r^3D^2V}\left|r^3x + \frac{r^2x^2}{2} - \frac{rx^3}{3} - \frac{x^4}{4}\right|_0^r$$

$$= \frac{3MGmk}{4r^3D^2V}\left(r^4 + \frac{r^4}{2} - \frac{r^4}{3} - \frac{r^4}{4}\right)$$

$$= \frac{3MGmk}{4r^3D^2V} \times \frac{11r^4}{12} = \frac{11MGmrk}{16D^2V}$$

(For the time being, the constant of integration is neglected).

Since the value of the gravitational constant G is determined experimentally, we can take that the operation by the constant of proportion, k, is also automatically accounted for in the value of G. Hence, we may neglect the factor k in the above equation.

Thus, $$W_2 = \frac{11MGmr}{16D^2V} \qquad (7/8)$$

Sum total (radial) work held in the body, from equations (7/7) and (7/8),

$$W = W_1 + W_2 = \frac{5MGmr}{16D^2V} + \frac{11MGmr}{16D^2V} = \frac{MGmr}{D^2V} \qquad (7/9)$$

Unequal momenta of forces about the centre of gravity (mass), of the free planet, cause body's simultaneous radial and spin motions. Equal momenta on either side of centre of gravity, together, cause radial motion of the planetary body. They act as single set of work (force) through the centre of gravity. Remaining one-sided momentum produces a couple about centre of gravity and causes spin motion of the body. Work, in the forward hemisphere and equal part of work in the rear hemisphere, together, produce planet's motion in perpendicular direction to its orbital path. This can also be understood as a shift in the location of planet's centre of gravity from its centre of mass, to the rear of the planetary body.

Total (radial) work acting through the centre of gravity,

$$W_g = \frac{5MGmr}{16D^2V} \times 2 = \frac{5MGmr}{8D^2V} \qquad (7/10)$$

This work, $5MGmr/8D^2V$, acts to produce planet's inertial motion towards the central body. Since no body can stay motionless in space, the factor V is always of positive value.

Remaining (radial) work, acting about the centre of mass of the body and producing spin motion of the planet,

$$W_s = \frac{MGmr}{D^2V} - \frac{5MGmr}{8D^2V} = \frac{3MGmr}{8D^2V} \qquad (7/11)$$

Magnitude of the central force and its components, W_g and W_s, also depend on the position of the planetary body in relation to the central body. In the relative position, considered above, direction of central force is perpendicular to planet's orbital path. There are points on the orbital path at which the planetary body experiences the central force in the same or opposite direction to the direction of its linear motion along the orbital path. At these points, magnitude of radial component of central force will be much higher and the magnitude of its spin component will be zero. At all other points in the orbital path, magnitudes of central force and its components will vary, cyclically, as the planetary body moves along its orbital path.

7.3.1. Magnitude of radial velocity:

Central body of a planetary system is very large compared to a planet. Therefore, it takes some time for the planet to move across the central body, in any tangential direction. They maintain common planes parallel to the radial direction during this time. As long as the common planes are present, they are under apparent gravitational attraction in that (perpendicular) direction. Apparent gravitational attraction, in any (perpendicular) direction, begins as soon as the forward part of planet comes in line with the central body and continues to be present as the planet advances or recedes in its orbital path, moving across the central body. Central force in any (perpendicular) direction ceases when the planet has fully crossed (moved ahead of) the central body in the corresponding tangential direction.

At the end of this time, all the (radial) work invested into the planet's matter field, for the production of its (perpendicular) radial velocity towards the central body, has been utilized (to change the direction of linear motion and to spin the planetary body). The planet ends its perpendicular acceleration in this direction. Actions of central force overlaps for near-by points on the orbital path. Radial displacement towards the central body, at every instant, is along different directions and it (in any radial direction) stops as soon as work introduced into the planetary body's matter field for motion in that particular direction is lost from the planet's matter field. Consequently, despite the continuous displacement towards the central body, a planet never reaches any nearer to the central body (disregarding variations required for eccentricity of the orbital path).

Equation (7/10) gives total (perpendicular) matter field-distortions held (or work) in the planet and producing its constant radial motion, 'u'. Unlike in the normal cases, where an external effort introduces matter field-distortions in a body, during the action of an effort, the case of planetary body and its matter field are different. This is because of the constant change in the direction of planetary body's linear motion. In any radial direction (to its instantaneous linear path), magnitude of (perpendicular) matter field distortion in the planetary body's matter field is constant. Therefore, there is no natural accelerating stage for the body during the introduction of radial matter field-distortions. The planetary body moves at a constant radial velocity (to its instantaneous linear path) along any radius of the orbit, where it is situated (this consideration lasts only for an instant). Line joining the centers of central and planetary bodies is considered as the radius of the orbital path. Accelerating stage, to develop this constant velocity, took place before the body came in the perpendicular direction considered.

Kinetic energy of a body, moving at constant speed, u,

$$KE = mu^2 / 2.$$

Mass of the planetary body 'm' is constant and its kinetic energy depends on its velocity.

Comparing these two; $\dfrac{mu^2}{2} = \dfrac{5MGmrk}{8D^2V}$ (from equation 7/10)

Radial velocity of the planet, $u = \sqrt{\dfrac{5MGrk}{4D^2V}}$ m/sec **(7/12)**

Although this radial velocity appears to be of constant magnitude (disregarding changes in distance, 'D' and absolute velocity, 'V'), it is being renewed at every instant. (Radial) work is lost from the planetary body and new (radial) work of equal magnitude is invested throughout the planetary body's matter field. Continuous loss of (radial) work keeps the radial velocity of the planetary body constant, despite continuous investment of (radial) work at the same rate. Investment of work produces a body's acceleration. Yet, in this case, final (radial) velocity of the planet is constant irrespective of its (radial) acceleration. This is because of the limitation on the ability of planetary body's matter field to store more (radial) work, than a constant maximum magnitude, imposed by its linear motion.

Planetary body starts to accelerate at a rate 'a' in its planes towards the central body, when it starts to cross a common plane with the central body. Acceleration in this direction ceases, when the whole body has crossed the common planes with the central body, in perpendicular direction. Thereafter its matter field is unable to store more work of this nature. Long before this time, it would have started similar actions in nearby planes also.

Constancy of radial velocity is mentioned only to emphasis the point that continuous action of central force in any particular direction does not change planet's radial velocity in that direction. However, due to gradual changes in the relative angular direction of central and planetary bodies, necessitated by a planet's displacement along its orbital path, magnitude of central force vary continuously. Depending on the present position of the planetary body in its orbital path, radial velocity of the planetary body varies continuously and cyclically.

7.4. Planetary orbits:

'Nebular hypothesis' suits the mechanism of formation of star systems in a galaxy. However, the planetary systems, like the solar system, are too small for such formations. Central body is too massive compared to the orbiting bodies, that it will assimilate any smaller body formed near it, as per the Nebular hypothesis. Hence, all orbiting bodies of a planetary system, like the solar system, have to come from outside the system. All parameters of a successful planetary body at the instant of its entry into its orbital path have to be equal to the parameters as if the body is in its stable orbital path about the central body.

Their entry speed should be high enough to prevent them from falling into the central body. Instead, they should orbit about the central body in an orbit determined by their entry speed and direction of approach. If the central body itself is under linear motion in any direction, which is usually the case, the direction of approach of the bodies trying to enter into orbits about the central body and become successful satellites are limited to certain angles of approach. All others will most probably be apparently attracted into the central body to fall into it or fly away from the central body. Entry of a body into planetary orbital path is a one-time process. There is no gradual development of an orbital path.

7.4.1. Linear motion of a rotating body:

Linear and rotary motions of a body are entirely separate phenomena. Each of them is produced by separate set of work-done on the body. However, each point on a linearly-moving rotating-body has its own path of resultant motion in space. Its motion and the path appear to be resultant of the linear and rotary motions of the body. In figure 7C, 'A' shows a rotating body that has no (absolute) linear motion. Centre point of the body 'O' may be assumed absolutely steady in space. Point P on its periphery traces a circular path, as shown by the circle in dashed line. Let the body develop a linear motion, as is shown by 'B' in the figure 7C and its centre of rotation moves from O_1 to O_2 at a constant speed, while the body turns through one revolution. Point P_1 on its periphery traces a loop as shown by the black curved line starting from P_1 and ending at P_2.

'C' in figure 7C shows the rotating body moving at a higher linear speed. Centre of rotation of the body moves linearly through a larger distance from O_1 to O_2, while the body turns through one revolution. Loop traced by a peripheral point becomes narrower as the linear speed increases, for the same rotary speed. 'C' shows the path of the peripheral point during one rotation of the body. Continuous loops in black from P_1 to P_n in 'D' shows a continues path traced by the peripheral point in space, while the centre of rotation of the rotating body moves linearly from O_1 to O_n along the line XX.

As the linear speed of the body increases in relation to its rotary speed, loops in the path of the peripheral point gradually becomes narrower until the loops altogether disappear at a stage. At this stage, the body's linear speed equals π times the radius of the rotating body (distance of the peripheral point from the centre of rotation of the body) during every rotation of the body. Black series of semi-circular paths 'E' in figure 7C shows the curved path traced by a peripheral point. Resultant path of the peripheral point consists of semi-circular curves with their convex sides in the same direction. The path starts from P_1 to P_n in 'E', while the centre of rotation of the rotating body moves linearly from O_1 to O_n along the line XX. As the linear speed of the body exceeds this value, no points in the body have motions in the reverse linear direction. All points in the body have displacements only in the

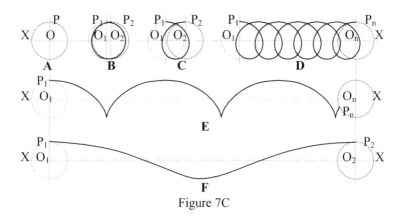

Figure 7C

forward direction. The requirements that points in the body on opposite sides of centre of rotation have motion in opposite directions are no more satisfied. No point in the body has circular/ elliptical path in space. All points in the body move in forward linear direction only. However, with respect to any point in the body, all other points in the body in its plane of rotation, appears to move in circular path around the point of reference.

As the linear speed of the rotating body is increased, circular path of the peripheral point expands to become a wavy path about the line of motion of the body's centre of rotation. The black curved line, 'F', in figure 7C, shows this path. Path of the peripheral point in space traces a wavy curve from P_1 to P_2, while the centre of rotation of the body moves from O_1 to O_2 along the line XX, during one rotation of the body. At lower linear speeds, difference between segments of curved path (on either side of the linear path) is large. As the linear speed of the body increases (for the same rotary speed), lower segment becomes larger and the difference between upper and lower segments of the curve reduces.

Although, depending on body's linear speed in relation to its rotary speed, the peripheral point traces curves of loops, semi-circular curves or wavy path in space, it still moves in a circle with respect to the centre of rotation of the body. Motion of the peripheral point in a circular path is apparent only to an observer situated at the centre of rotation of the rotating body. Circular path of the peripheral point, noticed by the observer, is an illusion due to the observer not considering his own linear motion in space. In fact, every point in the rotating body, moving in linear path, appears to move around every other point in the same body. This is a false impression, created by choosing a moving point as a reference. Every point has its own independent path in space. Other than when the rotating body has no linear motion, path of the peripheral point does not trace a closed geometrical figure in space.

A rotating body is moving in a linear path. Center of rotation of the body has a linear motion along a straight line, XX as shown in the figure 7C. For an observer, situated at one of its peripheral point, the centre of rotation of the body will appear to move around his location. He cannot observe his own true motion in space. He also cannot observe the linear motion of the rotating body (centre of rotation). Observed motion of the centre of rotation in a circular path around the peripheral point is an illusion.

Since the apparent motion of the peripheral point in circular path around the centre of rotation and the apparent motion of the centre of rotation in circular path around the peripheral point are only illusory motions, no true physical law can be based on them. Such illusory motions cannot be considered as proof of scientific laws. Observers, simultaneously situated at both these points will have apparent motions contrary to each other. None of them can observe the true motion of the points on the rotating body, in space. Real paths of any point on the linearly moving-rotating body can be viewed only from an external point. Origin of the frame of reference has to be outside the rotating body.

A rotating body's integrity keeps relative positions of its peripheral points with respect to its centre of rotation. Its integrity provides certain attachment between these points. All through their movements, distance between the centre of rotation and a peripheral point remains constant. Each of these points can appear to move in circular paths around the other point. Therefore, in any system of bodies, where distance between two bodies is always kept constant (by some means irrespective of bodies' motions) and where each of the bodies appears to move in circular path around each other, the above given explanations are valid. Apparent circular paths of planetary bodies (noticed by the observer) about their central bodies are not real and hence, it is not right to consider them as facts to develop scientific laws/theories.

7.4.2. Orbital motion:

A group of large bodies in space forms a planetary system. Bodies of this group move together along a median path, while individual bodies have independent relative motions within the group. The planetary system that includes the sun is the solar system. Path of each body in a planetary system is affected by the presence of all other bodies in it. For the time being, we may neglect effects on their paths by the presence of other bodies, outside the planetary system in space, as they are very small and the bodies are very far. There may also be smaller bodies called satellites in a planetary system. Satellites being very near to the planets, they form (sub) planetary system with their mother planet, within the larger planetary system. Largest body in the group has its path nearest to the median path and its path is least perturbed. This body acts as the leader of the group and it is the central body of the planetary system. All other bodies in the planetary system move along with the central body, while their paths are perturbed by the presence of all other bodies in the system. For the explanations below, we shall consider a planetary system containing a central body and one planetary body.

With reference to the planetary body, the central body appears to orbit around the planetary body and with reference to the central body, the planetary body appears to orbit around the central body. Disregarding the eccentricity of an orbit, distance between the central body and the planetary body remains constant. As shown in section 7.4.1; by these characteristic properties, a planetary system functions as a rotating body moving in linear path. Planet takes the place of a peripheral point and the central body takes the place of centre of rotation, in the above given explanation on the 'linear motion of a rotating body' in section 7.4.1.

A planetary system is essentially part of a galaxy. All stable galaxies are static in space (see section 7.7). Galaxies are rotating system of macro bodies. Hence, a planetary system in it traces a circular path around galactic centre. Median path of the planetary system is a very large circle around galactic centre. A small part of this very large circle is considered, here, as a straight line for the following explanations.

Actions of central force on a planet and a planet's orbital motion are independent of all other bodies, including the central body. Role of the central body or any other body in the vicinity is to limit the extent of 2D energy fields, acting on one side of the planet. Rest of all the actions are performed on the planet by the actions of 2D energy fields. Although a planetary body appears to move in orbital path around a central body, in reality, it has independent motion of its own. Apparent gravitational attraction towards the central body causes a planetary body's path to deviate from straight line, to move about and along with the central body in its motions. Due to the gravitational actions, orbiting bodies appear to influence the directions of each other's motion and create perturbations in their paths. Since a planet is a very small body compared to the central body, deviations in planet's path are more prominent. When these deviations are observed about a central body, which is assumed static in space, orbital path of the planet appears to be an orbit around the central body. This is the apparent orbit of the planet, which we observe in everyday life.

Similarly, relative to an assumed static planetary body, apparent direction of motion of the central body is around the planet. Few centuries back, when an earth-centered universe was in prominence, this apparent motion was considered true. Later as the science progressed, a heliocentric universe came into prominence. Earth, orbiting around the sun, is considered true in a heliocentric universe. Although we now know that, the sun is no more at the centre of universe, our view of planetary orbits in a heliocentric universe has not changed.

Orbital motion of a planet, around the central body, is apparent only with respect to the participating bodies. For apparent orbital motion, the body about which the observation is considered is the central body and the other body is the planet. With respect to an absolute reference (or an observer outside the planetary system and static in space), a planet does not orbit around the central body. Motion of the planet is wave-like along the central body's path, periodically moving to the front and to the rear of the central body as shown in figure 7D. In figure 7D, arrow in thick dotted gray line shows the path of the central body. This curved path, also, is wavy to a smaller extent, curving in the same directions as the

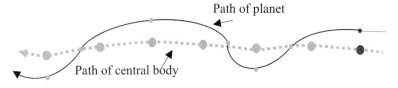

Figure 7D

path of planet. Arrow in black wavy-line shows planet's orbital path. Path of a satellite (not shown in the figure) of the planet is a wavy-line about planet's path. Central body and the planet are shown by black circles and their future positions are shown by grey circles. Asymmetry of parts of the curved paths about a centre-line, in the figure, is due difference in the scales used for 'X' and 'Y' axes in the figure.

In this sense, it can be seen that a planet (or a satellite) orbits around the center of the central body's curved path (galactic center) and the wave pattern in its path is caused by the presence of the central body. Such changes in the path of a free body may be attributed to perturbations caused by the presence of nearby bodies. These perturbations look like orbital motion, only when they are referred to a relatively small system of bodies. This argument can be carried further, to show that with respect to absolute reference there is no natural orbital motion around central bodies at all, except orbital motions of bodies around the (static) galactic centers.

Both, the planet and the central body move in the same direction about the same median path in space. Since the orbital motion is an apparent phenomenon, either of the bodies can be considered as the central and the other as its planet. Planetary laws are equally valid in either case. Although it is generally stated that the earth orbits the sun in eastward direction, it is equally valid to state, 'the sun orbits the earth in westward direction'. However, when more than two bodies are considered as a single system, it is more convenient to take the common and most prominent body as the central body and to take other bodies as planetary or satellite bodies.

Larger orbital path of a planet (and all bodies in a galaxy) is around the galactic centre. It is very large and contains many points of similar appearance in relation to the central body of a planetary system. Hence, we use a much smaller structure 'the apparent orbit' with unique reference points on it for all practical purposes. Apparent orbit is a small part of the larger orbital path, between two identical appearances of the central body, looking from the planet (e.g.: one solar year). It is an imaginary concept. As such, it has no logical basis. It depicts the appearance of a system, where it is assumed that the central body, by some imaginary mechanism (change of reference frame), is held stationary at the centre of the apparent orbit.

Although astronomers are aware of the real orbital path as wavy about the median path of the central body, most of them still consider apparent orbit as true orbital path of a planet. Since we consider instantaneous parameters of bodies, for all practical purposes of predictions, apparent orbits provide accurate results. However, it may be considered that planetary orbits around a central body were in consideration from the time of Copernicus' observations. The contemporary 'laws of planetary motions' by Kepler were formulated at a time, when even the heliocentric nature of solar system was not recognized and the phenomenon of gravitation was unknown. These laws were mathematical formulations of empirical parameters, observed by astronomers. They gave no logical reasons or theories, why the planets adhere to the observed apparent orbits around the sun, even in its static state. If the central body was a moving body, these laws did not adhere to simple mechanics. Having considered the earth as a moving body around the sun, the same laws were not suitable for moon's orbital motion, which was left out of the presumption. Unfortunately, modern physicists, even with the knowledge of the moving sun, adhere to the planetary laws (formulated for apparent planetary motion around a static sun), most devotedly. Acceptance of wavy-nature of planetary orbital paths can give simpler and logical explanations to many other puzzling problems in contemporary cosmology.

Only cause of actions within a stable planetary system is the central force, due to mutual apparent gravitational attraction, which accelerates a planet towards the centre of the apparent orbit – the central body. Currently, parameters of this action are mathematically manipulated to produce the required orbital motion around a central body that matches our observations. [In case of earth's apparent orbit around the sun, such mathematical treatments are also used to establish proofs of validity for Newton's "Laws of motion" and "Laws of universal gravitation"]. While doing this, (much greater) motions of the planetary body, before it became a planet and the motion or path of the central body are ignored. An apparent orbit is convenient to predict cyclic features that take place annually.

A non-circular apparent orbit has two reference points on it, periapse and apoapse. They are situated diametrically opposite in an apparent orbit. Apoapse is the point on the orbital path, where the planet is considered to be slowest and farthest from the central body and periapse is the point on the orbital path, where the planet is considered to be fastest and nearest to the central body. In real orbital motion of the planet, periapse is a point on its path, where it is nearest to the central body but the planet need not be fastest at this point. And apoapse is the point on its path, where it is farthest from the central body, but the planet need not be slowest at this point. In case of earth, these points are called perihelion and aphelion, respectively.

In an apparent orbit, direction of radial motion of the planet towards the central body is always (almost) perpendicular to its orbital path. In real orbital motion, it is not so. Radial motion of a planet is perpendicular to the orbital path only at datum points situated farthest and nearest to the galactic centre. At all other points on the orbital path, angle between radial motion and orbital path varies as the sine of relative angular position of the planet with respect to the central body and the median path.

Figure 7E compares the real path of an orbiting body and its apparent orbit for the duration of one apparent orbital period. Black central line is Central body's path. Wavy line is the path of the planet. Larger black circle shows the central body and the circles in dotted line show its future positions. Small black circle shows

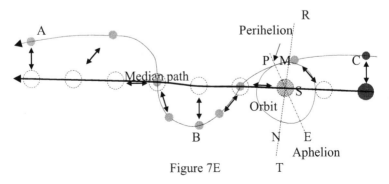

Figure 7E

the orbiting body and the grey circles show its future positions. Double-headed arrows show the central force due to gravitational apparent attraction between the bodies at various positions as they move along their paths. As a planet moves, its apparent orbit moves along with the central body.

Apparent orbit of the planetary body, when it is at position P with the central body at S, is shown by the oval figure in figure 7E. Apparent orbit's perihelion is at P and aphelion is at E. In real motion, highest and lowest linear speeds of the orbiting body occur, when it is at 90° away from the path of the central body, at M and B, respectively. All parameters of an apparent orbit and the orbiting motion are related to perihelion and aphelion. From its position at C, until B, the orbiting body is in front of the central body and hence it is retarded in its linear motion. From B to A, the orbiting body is behind the central body and hence it is accelerated in its linear motion. Line RST is the radial line connecting the central body to the centre of its curved path (galactic centre). Acceleration and deceleration of (the linear motion of) the planet change over at points M and B. These points are fixed relative to the path of the central body. Point M, on the outer side of central body's path may be called 'outer datum point' and point B (corresponding to point N on the apparent orbit), on the inner side of central body's path may be called 'inner datum point'. Datum points have only a vague relation to apparent orbit. A circular apparent orbit around a central body represents an ideal orbit about the central body, which may be called 'datum orbit'.

In figure 7F, 'P' is the planet at a point on its orbital path. XX

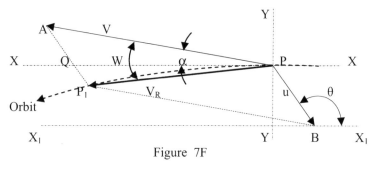

Figure 7F

is the tangent to the orbital path at 'P'. PA is the absolute linear motion (speed) of the planetary body, 'V', in magnitude and direction. Due to inertia, the planet tends to maintain the direction of its linear motion. Angle between 'V' and the orbital path of the body varies as the body moves in its curved path. Total angular displacement, 'W', produces the curvature of body's orbital motion. Direction of linear motion of the planetary body, moving in a curved path is deflected away from the tangent to the curved path.

At any instant, 'V' is deflected from XX by 'drifting rate', '-α' (clockwise deflection). Radial motion (velocity) due to the central force is PB = u. Direction of 'u' is towards the central body. Angle between 'u' and the tangent XX, at P is equal to the angular displacement, '+θ', of the planet in its orbit from a reference point, on the median path, X_1X_1, of the central and planetary bodies. PP_1 is the resultant motion of the planet, V_R, and it makes an angle 'W' with 'V'. Due to the action of the central force (radial motion 'u'), linear motion of the planetary body is deflected from PA to PP_1.

Angle between V and $V_R = \angle APP_1 = W$,

$\angle QPB = \angle PBX_1 = \theta$

Angle between V and u $= \angle APB = \theta + (-\alpha)$

$$\tan W = \frac{u \sin[\theta + (-\alpha)]}{V + u \cos[\theta + (-\alpha)]} \qquad (7/13)$$

'W' is the rate of angular deflection between present velocity, 'V', and resultant velocity, 'V_R'. It may be called the 'deflection rate'. 'α', the 'drifting rate' is the rate of angular deflection between the direction of present linear motion, V, and the tangent to the orbital path. In order to make the path curve towards the median path, resultant of deflection rate 'W' and drifting rate 'α' should be in the same direction as that of radial velocity 'u'. Vertical component (to the tangent XX) of present velocity 'V' is a real motion, substituting for the effect of the (presently) imaginary 'centrifugal force'. This part of real motion produces the drifting rate, 'α'.

Chapter 7. GENERAL

Inner solar System:

Figure 7G shows the real orbital paths of inner members of solar system. Planets and the sun, shown in the figure, are not to scale. Eccentricities of orbits are ignored. Relative positions of sun and the planets, shown on the right, are as on 3rd May 2002 [Reference: ESA Website]. Galactic centre is on the lower side and the solar system is depicted as rotating anti-clockwise around the galactic centre. Arrows at the ends of real orbital paths show the direction of motion of the sun and the planets. Small part of curved path of sun is shown as a straight line and its perturbations, caused by the planets, are not shown in the figure. Curved segments of planetary orbital paths, below the sun's path (on the side towards the galactic centre), appear narrower because of very small scale of distance used in 'X' axis compared to scale of distance used in 'Y' axis, in the figure.

It can be seen from the figure that the sun and planets move together along a common median path around the galactic centre. Presence of number of bodies in the system causes perturbation to the paths of all members. These perturbations, when observed with respect to any member of the system (which is presumed to be static) gives rise to the apparent orbits of closed geometrical (circular or elliptical) figures around the member, which is assumed to be static. The grey dashed lines around the sun show apparent orbits of the planets. Dim set of figures on the left shows the relative positions of the members of the system and their apparent orbits,

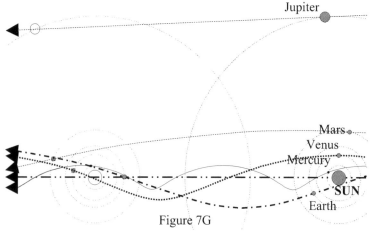

Figure 7G

five months later. Similar apparent orbits and the planetary system can be built about any member of the system. They all are imaginary.

Until last days of 16th century A.D., solar system was considered with earth based apparent orbits. Later on; due to popularity of Kepler's planetary laws and Newton's support to the same, the present system of sun-based apparent orbits for the solar system came into prominence. However, the imaginary nature of these orbital systems continues to be disregarded even after realizing the movement of the central body, the sun, in space.

Figure 7H shows the real orbital path of moon (a satellite) about the earth. Large black circle shows the sun and the straight arrow on the centre line shows small part of sun's path as a straight line. Small black circle shows the earth and the curved dotted line with arrow shows earth's real orbital path for five lunar months.

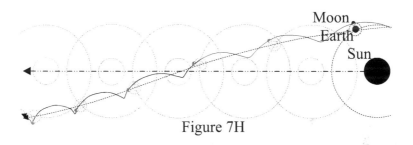

Figure 7H

Black wavy line shows the real orbital path of the moon. Lower parts of the real orbital path of the moon about the earth are shown narrower because of very small scale of distance chosen for the figure. Gray dashed circle around the sun shows the apparent orbit of earth. Black circle in dashed line around the earth shows the apparent orbit of moon around the earth. The figure is not according to any particular scale. Relative positions of sun, earth and moon are shown as for full moon days. Dim figures to the left show their relative positions and apparent orbits for subsequent full moon days. Eccentricities of the apparent orbits are not considered. Real orbital paths of all satellites about their corresponding planets are similar.

All bodies in the asteroid belts are planetary bodies with respect to the sun. Their real orbital paths, in the asteroid belt, are similar to planetary orbits about the sun.

Some planets are found to have many smaller bodies orbiting about them. These, when depicted in their apparent orbits, make picturesque rings about the planets. However, their real orbits are similar to the orbital paths of satellites about the planets. These bodies form a swarm around the planetary body and move along with the planetary body in its motions. Figure 7J shows the real orbital paths of these bodies about a planet. Figures on the right show the relative positions of the bodies and their apparent orbits. Large black circle shows the sun, small black circle shows the planet and the very small circles show three smaller bodies in the ring, situated in the same radial line from the planet. Large black arrow shows the path of the sun in a straight line. Black curved arrow shows the real orbital path of the planet. Curved arrows show the real orbital paths of the smaller particles in the rings. Dim figures on the left show the relative positions of the bodies and their apparent orbits after lapse of certain time.

Planetary bodies have lower rates of angular displacement

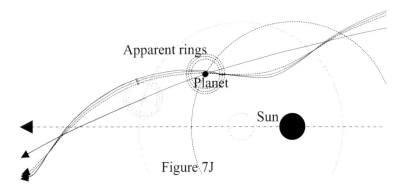

Figure 7J

with respect to their central bodies as their distance from the central body increases. Their orbital paths cross each other at different places in space. Due to their different angular speeds, there is a possibility for any two planets in a planetary system (or for any two satellites of a planet) to come very near to each other. At certain point of time in future, it is possible for any two planets in a planetary system (or for any two satellites of a planet) to collide into each other and destroy. In case of smaller bodies, forming rings about a planet, they have identical angular speed with respect to their parent body. This prevents them from colliding into each other during their motion along with their central body.

This swarm of smaller bodies about a planet also obeys all rules of planetary motions. Only those smaller bodies in the swarm, which are in (or very nearly in) the orbital plane of the planetary system can survive in the rings. All bodies, which do not conform to the planetary laws, will be automatically removed from the system by mutual collisions or rejections. Hence, the apparent rings about a planetary body are very thin and situated around planet's equator. Although, they are depicted as rotating around the planet, they also move along with the planet in its linear motion as shown by the curved lines in the figure 7J. Since their angular speed is the same, linear speeds of these bodies increase as the distance from the planet increases (within the escape velocity corresponding to the planet). Due to centrifugal action, caused by their angular speeds, larger (by matter content) bodies tend to distribute farther from the planetary body and smaller (by matter content) bodies remain nearer to the planet's surface.

7.4.3. Circular orbit:

Figure 7K shows the present and resultant orbital motions of a planetary body 'P', around a static central body. Present motions of the body comprises of its linear motion at an absolute speed 'V', represented by arrow PA and radial motion at an absolute speed 'u', represented by arrow PB.

Linear motion is provided by body's inertia and the radial motion is provided by the action of central force on the body. Arrow PP_1 represents the resultant motion of the body. For motion of a body in a circular orbital path, at any instant, its present linear speed 'V' is equal to its future linear speed 'V_R'. That is, at any instant, the resultant linear speed of the planetary body in its curved path is equal to its present speed. Line XX shows the tangent to the curved orbital path at present position of the body, at 'P'. In a circular orbit, deflection rate of direction of motion, 'W', is a

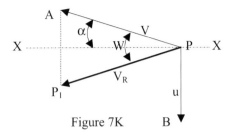

Figure 7K

constant. There is no angular acceleration. Hence, the drifting rate remains constant and equal to '-α' all around the path. If a negative drifting rate, '-α', can be maintained constant by external means or by natural process, the resultant linear motion of the body 'P', along its curved path, deflects at a constant rate and its magnitude remains a constant, equal to its present (instantaneous) linear speed.

In figure 7K; 'V' is the present (absolute linear) speed around a static central body. Direction of linear motion is deflected from the tangent XX at P by an angle '-α'. 'u' is the radial motion, perpendicular to the tangent, is the resultant speed of the planetary body and its direction is deflected from 'V' by an angle 'W'.

$$\angle APB = \tfrac{\pi}{2} + (-\alpha),$$

$$V_R^2 = V^2 + u^2 + 2VuCos\left[\tfrac{\pi}{2} + (-\alpha)\right]$$

$$V_R^2 = V^2 + u^2 - 2VuSin(-\alpha) \qquad (7/14)$$

For motion of the planet in a circular path; $V_R = V$.

Putting V in place of V_R in equation (7/14)

$$V^2 = V^2 + u^2 - 2VuSin(-\alpha),$$

$$u^2 = 2VuSin(-\alpha),$$

$$u = 2VSin(-\alpha) \qquad (7/15)$$

$$Sin(-\alpha) = \frac{u}{2V} \text{ or } (-\alpha) = Sin^{-1}\frac{u}{2V} \text{ rad} \qquad (7/16)$$

In a circular orbit, angle between the tangent at any point on it and the direction of radial motion of the planetary body due to central force is $\pi/4$ radians. Hence, $\theta = \pi/4$

From equation (7/13) $\tan W = \dfrac{u \sin[\theta + (-\alpha)]}{V + u \cos[\theta + (-\alpha)]}$

Putting, $\theta = \pi/4$, for circular orbit conditions;

$$\tan W = \dfrac{u \sin[\tfrac{\pi}{4} - \alpha]}{V + u \cos[\tfrac{\pi}{4} - \alpha]} = \dfrac{u \cos(-\alpha)}{V + u \sin(-\alpha)}$$

Putting the value of 'u' from equation (7/15) in equation (7/13),

$$\tan W = \dfrac{2V \sin(-\alpha)\cos(-\alpha)}{V - 2V \sin(-\alpha)\cos(-\alpha)}$$

$$= \dfrac{-2V \sin\alpha \times \cos\alpha}{V - 2V \sin^2\alpha}$$

$$= \dfrac{-\sin 2\alpha}{1 - 2\sin^2\alpha} = \dfrac{-\sin 2\alpha}{\cos 2\alpha}$$

$$W = \tan^{-1}\dfrac{-\sin 2\alpha}{\cos 2\alpha} = \tan^{-1}(-\tan 2\alpha) = -2\alpha \qquad (7/17)$$

Putting value of $-\alpha$ from equation (7/16),

$$W = -2\alpha = 2\sin^{-1}\dfrac{u}{2V}$$

$\sin \tfrac{W}{2} = u \div 2V$, $\quad \dfrac{W}{2} = \sin^{-1}\left(\dfrac{u}{2V}\right)$

$$W = 2\sin^{-1}\left(\dfrac{u}{2V}\right) \qquad (7/18)$$

For a circular orbital path around a static central body, where direction of 'W' is positive:

$\alpha = W \div 2$ in negative direction.   **(7/19)**

This is the condition required for a circular orbital path of a planet around a static central body or circular parts of other orbital paths. Resultant linear speed (in absolute terms) of the planet, along the curved path, remains a constant equal to its present (instantaneous) linear speed. Angular speed of the planet (deflection rate) is equal to twice the drifting rate (in opposite angular direction) and it is a constant. Drifting rate of a planet, required to achieve a circular orbit (in this case) is less than the angle of contingence at the point of initial entry, on the datum orbit (clockwise from the tangent at P). It is precisely equal to half the rate of deflection rate produced by the central force at the current distance between the bodies. Hence, the planet entering its orbital path is required to initially approach the entry point 'P' from within the datum orbit. These conditions can be met only in cases, where the orbit is formed around a static central body.

All natural planetary bodies are much smaller than the central bodies of their planetary systems and they approach their orbits from outside their datum orbits. In real orbital motion, a planet in its orbital motion, traces segments of curved paths on either sides of its median path (which is also the median path of the central body). A circular orbital path requires semi-circular paths on either sides of median path. Due to constantly changing relative direction of central force, it is also impossible to maintain constant angular speed by a planet about a moving central body. Consequently, natural planetary bodies cannot have circular orbits around their central bodies (which are moving).

Exceptions to the above are probable (hypothetical) cases of static binary systems or other planetary systems formed by explosion of a static parent body, where the planets are thrown away from a static central body to enter their orbits from within datum orbit. Since, no body except galaxies, can remain static in space, these are only a theoretical consideration.

Circular orbit is a critical condition. Parameters of a planetary body (maintained in a circular orbit by external means, in addition to the central force) are very precise. Once in the orbit, the drifting rate can be easily changed by external factors. Changes in the matter contents of the planetary and central bodies or their speeds due to external influence are bound to affect the stability of a circular orbit, by changes in the drifting rate. Collision with debris in space or even uneven distribution of mass of the bodies can influence the state of a circular orbit. It should also be noted that no bodies, smaller than a galaxy, can remain static in space.

To form a circular apparent orbit about a central body, parameters of an orbiting planetary body should satisfy the equation (7/18), $W=2\sin^{-1}(u/2V)$, at every point on its orbital path. All factors in the equation remain constants. Apparent circular orbit is the smallest apparent orbit. It is the datum orbit of the planetary body for its present parameters. This equation is also applicable to circular parts of non-circular orbits.

7.4.4. Elliptical Orbit:

Variations in the parameters of an orbiting planetary body change its datum orbit. Consequently, even if a planetary body was in an apparent circular orbit for any length of time, its datum orbit will change from circular to non-circular apparent orbit, on variation of any parameter of the body. Such a change or a difference in the drifting rate changes the shape of the orbital path. Non-circular apparent orbits are based on the datum orbit of the planetary body. A deformed circular datum orbit becomes non-circular apparent orbit of the planetary body. Deformation of the datum orbit takes place with respect to two points (mid-points) that are on diametrically opposite sides on the apparent orbit. Either forward or rearward part of non-circular apparent orbit is placed within and the other part is placed outside the circular datum orbit.

Since a planet moves in a non-circular apparent orbital path, tangent to a point on the orbit is not necessarily perpendicular to the radius of the orbital path, along which the central force is acting. Nevertheless, there are two points that lie on the apparent orbital path, at which the conditions required for circular orbits are satisfied. At these points, direction of radial motion is perpendicular to the tangents to the curved path and the direction of change in the length of apparent orbit's radius reverses. If the radius of the apparent orbit was increasing before the planet reaches the point, after crossing this point, it will gradually decrease until the planetary body reaches a similar point on the diametrically opposite side of the apparent orbit. At this point, direction of change in the distance between the bodies reverses and the radius of the apparent orbit gradually increases until the planetary body reaches the original point on the apparent orbit. Periodic changes in the length of radius of the apparent orbit about a mean value sustain the stable non-circular apparent orbital path. Points, where these reversals occur are perihelion and aphelion, at which the bodies are nearest or farthest from the central body. Other reference points on the orbital path are the outer datum point and the inner datum point, where the planet attains highest and lowest linear speeds.

In case of real motion, an orbital path is not around the central body but it oscillates about a common median path, shared by the central body. Angular speed of an orbiting planetary body does not correspond to motion in circular or elliptical paths. Deflection rate of orbital path from the median path is limited, alternating on both sides. Whereas, a total deflection of 2π radians, in the same angular direction, is required for every apparent orbit.

Linear motion of a planet is accelerated up to the outer datum point, when the planet moves to the front of the central body or it is decelerated up to the inner datum point, where the planet falls behind the central body. Highest and lowest speeds of the planet occur at these datum points. These datum points need not coincide with either perihelion or aphelion of the apparent orbit, assumed above. Datum points of orbital path are situated on radial lines of the galactic radius (line perpendicular to the median path of planetary system) and passing through centers of both the central and planetary bodies. Points of perihelion or aphelion indicate the points, on the path of the planet, where the planet is nearest or farthest from the central body, respectively. They are not related to the motions of the orbiting planetary body. Perihelion and aphelion of an orbit may be displaced along the orbital path without affecting other parameters (excepting the planetary spin speed) of orbital motion. However, points at which the acceleration/deceleration change-over in the linear motion of the planetary body takes place, are fixed with respect to the median path of the central body and depend on the relative position of the planet with respect to the central body.

Part A of figure 7L shows planet 'P' at the perihelion of an orbit and part B of figure 7L shows the planet 'P' in the same orbit at its aphelion. 'V' is the present speed, 'u' is the radial speed and 'V_R' is the resultant speed of the planetary body. Drifting rates are equal to '-α'. 'W' is the deflection rates of the resultant motion. Changes due to body's angular acceleration / deceleration add to the drifting and deflection rates. Motions of the planet at perihelion and aphelion of its non-circular orbit exhibit properties of circular orbit. Angles between direction of radial speed 'u' and tangents at perihelion and aphelion are 90° each.

$$\tan W = \frac{u \sin[\theta + (-\alpha)]}{V + u \cos[\theta + (-\alpha)]} \quad \text{by equation (7/13)}$$

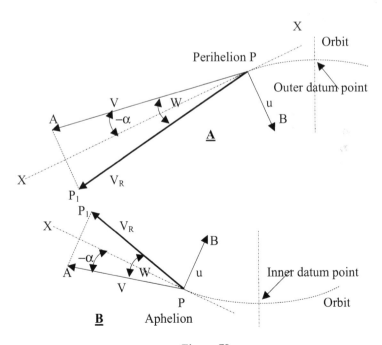

Figure 7L

Putting drifting rate, $-\alpha = W/2$ and $\theta = \pi/2$ in equation (7/13);

$$\tan W = \frac{u \cos \frac{W}{2}}{V - u \sin \frac{W}{2}} \quad (7/20)$$

$$V \sin W - u \sin W \sin \tfrac{W}{2} - u \cos W \cos \tfrac{W}{2} = 0,$$

$$V \sin W - u \cos\left(W - \tfrac{W}{2}\right) = 0$$

$$2V \sin \tfrac{W}{2} \cos \tfrac{W}{2} - u \cos \tfrac{W}{2} = 0,$$

$$2V \sin \tfrac{W}{2} - u = 0, \quad \sin \tfrac{W}{2} = u \div 2V$$

$$W = 2 \sin^{-1} \frac{u}{2V} \quad (7/21)$$

Equation (7/21) for elliptical orbit, at perihelion and aphelion, is same as equation (7/18) for a circular orbit. Since the curvature of the path varies continuously, this circular behavior of the orbital path lasts only for an instant. As soon as the points of perihelion or aphelion are passed, planet will pursue its non-circular path. Drifting rate at perihelion, α_{peri}, is half of deflection rate, W_{peri}, and at aphelion; drifting rate, α_{aphe}, is half of deflection rate, W_{aphe}, in magnitudes. As shown in figure 7K, directions of drifting rates with respect to radial motion are always the same. Included angles between radial motion 'u' and direction of approach are of the same sense. In a non-circular orbit; equation (7/13) gives the deflection rate. Equation (7/21) for circular orbit is applicable to a non-circular orbit at its perihelion and aphelion, where circular orbital conditions exist.

Radial velocity of the planet,

$$u = \sqrt{\frac{5MGrk}{4D^2V}} \quad \text{m/sec by equation} \quad (7/12)$$

Assuming the value of constant k in the equation (7/12) is integrated into the gravitational constant G and putting value of 'u' from equation (7/12) in equation (7/21);

At perihelion or aphelion, magnitudes of deflection rate;

$$W = 2\sin^{-1}\frac{\sqrt{\dfrac{5MGr}{4D^2V}}}{2V} = 2\sin^{-1}\sqrt{\frac{5MGr}{16D^2V^3}} \qquad (7/22)$$

Deflection rates 'W' for perihelion and for aphelion are in opposite directions. If the planetary body has entered its datum orbit from within (with negative drifting rate), it will move towards its aphelion, where conditions of circular orbit take place. If the planetary body has entered its datum orbit from outside (with positive drifting rate), it will move towards its perihelion, where conditions of circular orbit take place. Similar conditions (as required for circular orbit) should also repeat at a point 180° (in the apparent orbit) away from the perihelion / aphelion, as the case may be.

A stable orbit can be formed only if these conditions are met. At the perihelion, direction of change in the distance between the central and planetary bodies reverses. The planet moves towards its aphelion, where condition for circular orbit is fulfilled once again. At mid-points between perihelion and aphelion, deflection rate, 'W' becomes equal to drifting rate '-α'. Resultant angular speed of the planetary body becomes zero. For an instant, planet moves in a straight line. Direction of linear motion of the planetary body is tangential to the orbital path at these points but direction of radial motion is not perpendicular to the tangent. Therefore, conditions for circular orbit are not fulfilled. Tangents at these points are not parallel to the major axis of the apparent orbit. In a theoretical elliptical path, tangents at the ends its minor axis of the ellipse are parallel to major axis of the ellipse. In this case, they are not so. Hence, apparent orbital path of a planet is oval (with its narrower end towards the aphelion) in shape rather than an elliptical. Unlike an ellipse that has two foci; apparent orbital path has only one focus. However, due to very small eccentricity of the apparent orbits, we come across in nature and simpler mathematical treatments; they are usually considered as elliptical or circular. [Or it could be due to firm belief in Kepler's planetary laws, which categorically states that a planetary orbit is elliptical in shape.]

Present planetary laws are formed for apparent elliptical apparent orbits.

After the mid-point, between perihelion and aphelion, angular difference between the radial and linear motions diminishes. When the planetary body has moved from this mid-point by an angular displacement equal to the deflection of perihelion or aphelion from the datum points, linear and radial motions of the planetary body become co-linear. At this point, deflection rates of the planetary body, due to central force, becomes zero. However, the planetary body continues to move in its curved orbital path under the influence of drifting rate, which continues to decrease in magnitude. Once, the mid-point is passed, direction of angular difference between linear and radial motion reverses. Deflection rate 'W' and drifting rate 'α', both are in the same direction for a short while until drifting rate 'α' changes its sense. The planet angularly accelerates till it reaches another point, where (conditions for circular orbital path are fulfilled) deflection rate 'W' and drifting rate 'α' are in opposite directions and magnitude of deflection rate 'W' is twice that of drifting rate 'α'. This point is the aphelion of the orbit. Thereafter, similar processes continue to sustain the stable orbital motion.

Resultant orbital angular speed at perihelion

$$= W_{peri} - \alpha_{peri} = \omega_{peri}$$

Resultant orbital angular speed at mid-points $= W_{mid} - \alpha_{mid} = 0$

Resultant orbital angular speed at aphelion

$$= W_{aphe} - \alpha_{aphe} = \omega_{aphe}$$

Time to move from perihelion to aphelion $= T \div 2$.

Where, T is orbital time period.

Resultant orbital angular speed decreases from ω_{aperi} at perihelion to zero at mid-point, increases from zero at mid-point to ω_{aphe} at aphelion, decreases from ω_{aaphe} at aphelion to zero at mid-point and increases from zero at mid-point to ω_{aperi} at perihelion.

Taking the variation in the angular speed to be uniform;
Total difference between angular speeds at perihelion and at
aphelion $= \omega_{peri} + \omega_{aphe}$
Orbital angular acceleration/deceleration

$$= (\omega_{peri} + \omega_{aphe}) \div \tfrac{T}{2} \qquad (7/23)$$

Location of 'perihelion' or 'aphelion' of an orbital path depends on the location of point of entry on the datum orbit and the drifting rate at the time of entry. For appropriate drifting rate, the point of entry can also be the perihelion or aphelion of the orbit. Location of perihelion/aphelion can shift later due to external influences, whereas, the datum points remain at their relative positions with respect to the central body.

Limits of angular speed during entry:

Real orbit of a planet is a path in space, whose parameters are related to the central body. It is improbable for planets to be borne in their orbits. They have to come to their orbits from space, away from orbital path. For a smooth transition, from their motion outside the orbit into the orbital path, all their parameters of motion at the point of entry should be same, as if they were moving in a stable orbit at that point. There is no possibility of gradual stabilization of an orbital path, as is envisaged at present. Gravitational action, subscribing to the central force, is active on planetary bodies even when they are very far from orbital path. Hence, parameters of planetary bodies' linear motions are modified continuously, even before they enter their orbits. A planetary body enters its orbital path in near-tangential direction, subject to the following limits. Bodies, approaching the point of entry into the datum orbit outside certain limits of their angular speeds, are unable to form stable orbits. As the magnitude of drifting rate, 'α', approaches a limit in negative (clockwise) direction, magnitude of deflection rate, 'W', becomes insufficient to overcome the drifting rate and the direction of the resultant motion, V_R, becomes parallel to the tangent. Such a body is not able to form an orbit about the central body.

When V_R is along the tangent in figure 7L;

Drifting rate, -α, is equal to deflection rate W and $\frac{u}{V} = -\sin\alpha$

$$-\alpha = \sin^{-1}\frac{u}{V}$$

This is the lower limit of drifting rate, 'α', at the point of entry (from within the datum orbit) for bodies, which may form successful orbits about a central body. Bodies, approaching the datum orbit from within, with equal or higher (negative) drifting rate than this value will fly away from the central body.

Equation (7/16), for the value of $-\alpha = \sin^{-1}\frac{u}{2V}$, gives the condition required for an orbiting planetary body to have a perihelion and an aphelion in its apparent orbital path. This equation should be satisfied two times in every completed apparent orbit. If the planetary body is entering its datum orbit from outside, by the time it reaches its perihelion, drifting rate of the planetary body attains a value of $\alpha = \sin^{-1}\frac{u}{2V}$. As long as this value is not reached, the planetary body will continue to move towards the perihelion. That is, the distance between the central and planetary bodies continues to reduce. If the drifting rate exceeds the value, $\alpha = \sin^{-1}\frac{u}{2V}$, the orbiting planetary body will move towards the central body at a higher rate and spiral down into it, without ever attaining the condition required for perihelion. Even if the orbiting planetary body is to enter the datum orbit at the point of perihelion, its drifting rate should not exceed this limit. Thus, $\alpha = \sin^{-1}\frac{u}{2V}$ is the outer limit of drifting rate during entry (for a planetary body entering the datum orbit from outside) for a successful orbit about its central body.

Considering the above limits together; to form a stable orbital motion about a central body, a planet has to enter its datum orbit with value of drifting rate between $\sin^{-1}(u \div 2V)$ and $-\sin^{-1}(u \div 2V)$ at the point of entry. Limit between $-\sin^{-1}(u \div 2V)$ and zero is for those bodies approaching from inside the datum orbit.

Bodies with drifting rate between $\sin^{-1}(u \div V)$ and $-\sin^{-1}(u \div 2V)$ will have their aphelion in front of the entry point. When the drifting rate is equal to the critical value of $-\sin^{-1}(u \div 2V)$, the planetary body will trace semi-circular orbits or either sides of the median path.

Bodies with drifting rates between $-\sin^{-1}(u \div 2V)$ and $\sin^{-1}(u \div 2V)$ have their perihelion in front of their point of entry. Limits between zero and $\sin^{-1}(u \div 2V)$ are for those bodies approaching from outside the datum orbit. From the point of entry, they can move only towards their perihelion.

These stringent restrictions, in conjunction with restrictions on the magnitude of angle of entry (direction of approach as explained in the next sub-section on 'Orbits about a moving central body'), considerably lowers the number of bodies, which are able to form stable orbits and prevents profusion of planetary bodies about a central body.

Curvature of orbital path and the tangential speed of a planet, at any time, depend on its location on the orbital path. Center of curvature at any point on the orbit is the focus of the orbital path. In real-motion, centre of curvatures for orbital motion on either side of median path, lie on the opposite sides of the median path. Curvature is zero at mid-points and increases as the planetary body moves towards the aphelion or perihelion. However, while considering the apparent orbit, the imaginary curved orbital path is assumed to close-in on itself to provide circular / elliptical nature.

Orbits about a moving central body:

It is unlikely that bodies of considerable sizes move away from a central body to enter into orbital path about it. All larger bodies like the planets in a planetary system have to come from outside a planetary system. Planets may enter into their orbits in any direction around a static central body. However, if the central body is moving as is the case with all free natural bodies in space, directions of approach and other parameters of the planetary bodies are restricted. Following description is about planetary bodies approaching the central body from outside their datum orbits. To make the explanation simpler, an apparent orbit is used as a reference.

All large bodies, in space, move at very high speed. (Sun moves in a circular path around the galactic center at a relative speed of about 250000 m/sec, much greater than the relative speed of earth with respect to the sun, which is about 30000 m/sec, in its orbit). Relative speed between the central body and a planet, trying to enter into an orbit, depends on the relative directions of their motions, with the limitation that the linear components of the motions of both bodies are always in the identical directions.

Planetary bodies, approaching in opposite direction to the motion of the central body will find the interaction due to central force enhancing their present speed. Relative speed of the bodies will become too large for them to form a planetary system. Consequently, no planet that is approaching the central body in opposite direction to the direction to its own motion can enter into a successful orbit about that central body. Similarly, bodies approaching from the sides (all around) will be left far behind the central body. Such bodies have very little or no motion in the direction of motion of the central body. Hence, their relative speeds are too large. They cannot form stable orbits about the central body. In order to enter into a successful orbit, a planetary body has to approach the central body from the rear and nearly in its (future) orbital plane. During such an approach, relative speed of the planetary body varies only by a small fraction with respect to the central body's absolute speed.

A central body usually has a curved path as shown by line NOM in figure 7M, where direction of approach of a planetary body is shown with respect to its apparent orbit. A planet approaching its datum orbit from the inner side of this curved path, NOM, will find that it has an additional relative motion away from the galactic center. This is produced by the curvature of central body's path. Planet's drifting rate is enhanced by the curvature of central body's path. Additional relative motion will also enhance the radial motion produced by the central force. These factors prevent a body, approaching from the concave side of the central body's path to enter into a successful orbit (about the central body that itself is moving in a curved path).

The above-mentioned factors leave only a small conical window, shown by APEC in figure 7M (width of the window shown is highly exaggerated), through which a planetary body may enter into successful orbit about a central body. This window is on the outer (convex) side of central body's curved path, NOM, and to the rear of the central body. The window, in 3D space, is somewhat conical in shape, with its apex towards the outer datum point in the orbital path. Girth of the cone restricts entry to the bodies, whose orbital plane can be gradually stabilized into central body's orbital plane. All planets, entering into successful orbit, enter through this window, which is further restricted by the limits of drifting rate. Therefore, there are no planets orbiting in opposite direction to central body's own orbital direction or having its orbital plane too far from central body's orbital plane. Direction of apparent orbital motion of a planet is the same as the direction of central body's orbital motion about the galactic center. Thus, all orbiting bodies in a planetary system move in the same apparent angular directions and in planes not much different from central body's orbital plane.

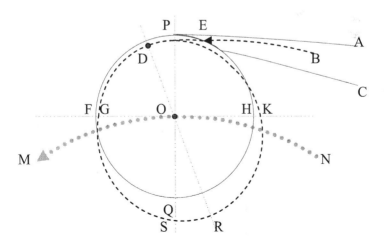

Figure 7M

As shown in figure 7M, O is the center of apparent orbit. Black circle at O is the central body and the black circle at D is the planet at its perihelion. PFQH is the datum orbit corresponding to planetary body's parameters. P and S are the datum points, where highest and lowest linear speeds of the orbiting planetary body occur. Orbiting planetary body may enter into the datum orbit anywhere through the conical window shown by the region between A and C. Position of perihelion of the orbit, D, depends on the point of initial entry, E, and the drifting rate. Greater the drifting rate, farther from the point of entry is the perihelion.

Only those planetary bodies, whose drifting rate, at the time of entry into datum orbit, are within the limits; $\operatorname{Sin}^{-1}(u \div 2V) < \alpha < -\operatorname{Sin}^{-1}(u \div 2V)$ and whose direction of entry is through the permitted conical window can produce a stable orbit about a central body that is itself moving in a curved path. Changes in the mass or speed of an orbiting planetary body may change the size of its orbit, not its eccentricity and angular position. In order to change the eccentricity or angular position of an orbit, an external effort (force) has to be applied on the orbiting planetary body. The external effort has to deflect the planetary body from its course and change the deflection rate of linear motion, V, by an effective change to the drifting rate, 'α', while in the orbit.

Figure 7N shows the entry zone for planets on their real orbital path. NOM is the path of the central body. KDRK is the orbital path of the planet and APC shows the entry zone. This figure may be compared with figure 7M. When figure 7N is used to depict the real orbital path of a planet, instead of the apparent orbital path, as

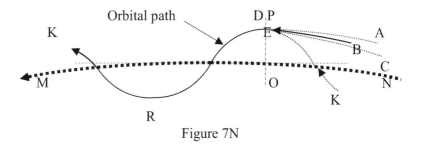

Figure 7N

is used presently, it is clearly evident that a planetary body approaching a central body from any quarters other than through the entry window APC cannot enter into successful orbit about the central body. Entry window APC is outside the apparent orbit of the planetary body. Hence, all bodies entering into stable orbit about a central body have to come from outside the planetary system.

Unlike the planets, comets have very large and highly eccentric apparent orbits. Their periapse occur within central body's path and highest absolute orbital speed occurs nearer to their apoapse. Hence, it may be deduced that they enter their orbit from within their datum orbit.

7.4.5. Anomalies in planetary orbits:

Ideal cases of orbital motions of an orbiting body and its central body are considered in the above explanations. However, in space, there are many other bodies in the vicinity of a planetary body and none of them are static in space. Orbital motion of a planetary body is affected by the presence of nearby bodies and their motions. These effects usually produce anomalies in the ideal orbital motions of a planetary body with respect to its own central body. Mostly, a central body may have a number of planets, each one with its own satellites. All these bodies affect the orbital paths of all other bodies in the planetary system.

Apparent loss of orbital motion of a planet:

While a planet is performing its orbital motion about the central body, it is also orbiting around the centre of central body's orbital centre (the galactic centre). Consider a planetary system as a single unit (of revolving bodies about a central body) that is orbiting around the galactic centre. Galaxies are static spinning bodies and they have no orbital motion about any other body. [During their formations and stabilizations, they may have relative linear motions towards or away from near-by galaxies.] By the time the central body completes an orbit around the galactic centre, every planet in the unit apparently loses one apparent orbit each, about the central body. This provides an apparent loss of orbital motion to the planet. A planet apparently loses part of its orbital motion, about the central body, at a constant rate.

Precession due to eccentricity:

If there is a large difference in the matter contents of the central body and a planet that has highly eccentric orbit, difference between linear speeds of the orbiting planetary body during its linear acceleration and deceleration stages is considerably large (magnitude of action of an external effort depends on the linear speed of a body, in the direction of effort). This speed difference influences the central force's actions on the planet. Such actions can apparently rotate (precess) an apparent orbit without changing its shape or size. Rotation of the apparent orbit means a forward/rearward shifting of the real orbital path of the planet in relation to central body.

In real motions, both the central and the planetary bodies move about a median path. Perturbations of planet's path are more apparent and it appears to be its orbital motion about the central body. While a planet is moving along with the central body, it is in front of the central body for half the orbital period (during its motion from outer datum point to inner datum point) and it is behind the central body for the next half of the orbital period (during its motion from inner datum point to outer datum point). When the planet is in front of the central body, the central force acts to decelerate the planet and when it is behind the central body, the central force tends to accelerate the planet. (Actions on the central body are of opposite nature).

Inertial actions are slower and hence, less effective in the direction of motion of a planet. Hence, it takes longer for the planet to traverse the path from inner datum point to outer datum point compared to the other half from outer datum point to inner datum point. Longer time-period causes larger radial displacement of the planetary body from the central body, taking it nearer to the median path (during its travel from inner datum point to outer datum point). Planet's perpendicular distance from median path during this period becomes shorter than its distance in opposite direction during the next half of the orbit. This difference provides a resultant displacement of the orbital path towards the center of central body's path. The planet, in its path is moved nearer to the center of central body's path (galactic centre).

In an apparent oval orbit, there is only one point that is nearest to its focus. This point is the perihelion. If the orbital path is displaced, such that another point in the orbit comes nearest to the focus, it is as if the perihelion of the apparent orbit has shifted to the new point. As different points on the apparent orbit come nearest to the central body, on successive apparent orbits, apparent orbit's perihelion shifts along the apparent orbit and the apparent orbit of the planetary body appears to rotate about the central body, in space. This precession is an illusion provided by displacement of 'point of nearest approach' of the bodies. In reality, no changes take place in the relative motions of the planetary body, except that the point in space, at which these bodies come nearest shifts along the perturbed path of the planet. Magnitudes of the perturbations or their time-periods are not affected.

If orbital motion is considered with respect to the planet (considering the central body as orbiting the planet), orbit of the central body will appear to precess in the opposite direction. Apparent orbits of all planetary bodies with highly eccentric orbits have appreciable precession about their central bodies. In case of a central body, having two or more such planetary bodies in the apparent orbits around it, apparent orbit of the central body has different rates of precessions simultaneously; a different rate of precession with respect to each of the planets. This is an impossible situation, if the precession is linked to real motion of the body. Direction of this precession is the same as the direction of motion of the planet in its orbit.

Precession due to central body's path:

Apparent planetary orbits are imaginary paths in space around a static central body. In reality, the central body moves in a curved path in space. Consequently, as the central body moves in its curved path, the median path of a planet is required follow this curved path. At the same time, inertia of the planetary body tends to move it in a straight line. Due to their relative motion (caused by the curvature of central body's path), median path of the planet tends to shift outward from the center of central body's path and with respect to the median path of the central body. In reality, it is the central body's path, which shifts in space with respect to the median path of the planet. Under inertial actions, median path of the planet tries to be straight, while central body's path curves. A shift in the median path of the planet introduces corresponding shift in the central body's path also. Real inward curvature of central body's path appears to be an outward shift of planetary orbital path with respect to the central body.

Consequently, the planet is on the concave side of the median path of planetary system for shorter time compared to its existence on the convex side of the median path. Additional radial displacement in perpendicular direction from the median path, provided by the central force, brings the median path back to its original state. By doing so, the median path of the planet is shifted (curved) towards the center of the central body's path. This displacement of the planetary orbital path is in addition to its normal displacement required for planet's orbital motion about the central body. Shift of the median path also shifts the 'point of the closest approach of central and planetary bodies' in opposite direction, away from the galactic center. This relative motion creates another type of apparent precession of the orbital path. Direction of this precession is in a direction opposite to the orbital motion of the planet. This precession is an illusion provided by relative inward displacement of median path of the planet with respect to the central body's path (caused by the curvature of central body's path).

Two types of precessions, mentioned above, are in opposite directions. In cases of apparent orbits with low eccentricity, they, more or less compensate each other. No resultant precession of a planetary apparent orbit may be noticed over extended periods. In cases of apparent orbits with higher eccentricity, precession caused by the orbital eccentricity may be large enough to produce observable resultant precession of the apparent orbit in forward direction. If the eccentricity of the apparent orbit approaches zero, precession caused by the curvature of central body's curved path may become apparent over extended period as a rotation of the apparent orbit in rearward direction.

Perturbations caused by collisions:

Wandering bodies in open space may fall into planetary bodies or into central body during their attempt to form orbits about either of the bodies. A foreign body falling-in, brings-in additional matter content and work associated about it. Such additions are likely to modify orbital parameters of central and planetary bodies. Over extended period of time in space, such modifications of orbital properties of planets are very probable. Bodies in the neighborhood of planetary bodies also influence their orbital motions. These changes may be either temporary or permanent. An accurate picture of planetary orbital motion can be developed only when all other bodies in space are considered.

7.4.6. Electronic orbits:

From the largest of the stars to the smallest of fundamental particles, if a body has to orbit about another, it has to do so under the same principle and such motions should obey the same physical laws. Any matter body with sufficiently large linear speed may orbit about a moving central body of (comparatively) appreciable matter content and linear speed. Electrons are matter bodies with definite physical structures. In their stable orbital motion, central force (provided by apparent gravitational attraction) is the only effort acting between them and the nuclei of atoms, to which they are attached. Electrons, in the atoms, keep their attachments with the nuclei under the central force provided by apparent gravitational attraction. Orbits of atomic electrons are located in a region, where the electromagnetic or nuclear manifestations of field force have no effects. [Various forces mentioned are different manifestations of only one type of force in nature. It is the nature of action of a force that makes its class. Hence, we may say that the gravitational nature of force is prominent and nuclear and electromagnetic nature of the force is absent about an electron in stable orbit about its nucleus]. During unstable periods, electromagnetic forces develop between nucleus and electrons to stabilize their orbital motions. These stabilizing forces determine the size, angular and linear speeds, eccentricity and location of the electronic orbits. It is the orbital motion, (in addition to the repulsive electric field forces during unstable periods), which prevents an electron from spiraling down into atom's nucleus or flying away from it.

Since the electrons are too small and curved paths of their central bodies are within the electronic orbits, electronic orbits in stationery atoms appear to be real orbital motion around a moving central body. An electronic orbit is the nearest, a real orbit can resemble apparent orbital motion. Electronic orbits remain on the same side (outside) of the central body's path and the deflection and drifting rates do not change directions. Electrons in the atoms also have apparent orbital motion about their nuclei. In these cases, the central bodies (corresponding deuteron paired to each of the electrons) are anchored to the nucleus and hence they are unable to have independent orbital motion of their own, other than that provided by spin motion of the nucleus. All planetary laws are applicable to orbiting electrons, with modification as required due to controlled motion of their central body and their orbits being placed outside the central body's path.

7.5. Planetary spin:

All bodies in a planetary system tend to spin about their diameters, perpendicular to the orbital plane. Part of the central force on the orbiting body causes the body to spin in the angular direction opposite to its apparent orbital motion. During its orbital motion, a planetary body alternates its location on either side of the median path, every half-orbital period. Spin motions, produced on either side of the median path are in opposite directions and they tend to compensate each other. Only the resultant-spin motion, over a long period, shows up with the body.

Bodies of a planetary system may or may not have spin motions (on their own) during their entry into the union. Once the planetary body has settled into an orbital path about the central body, during its motion in the orbital path, it develops a spin motion about one of its diameters perpendicular to the apparent orbital plane. If the planetary body was already spinning, before it entered into the orbit, its spin speed and the direction of spin are modified gradually, as required by the inertial actions to suit the present conditions. Development of spin motion of the central body is also similar. In this case, the smaller body acts as the central body and the larger body acts as the orbiting planetary body. If there are more planets in the planetary system, each planet will contribute to spin the central body, corresponding to its parameters.

Central force between a planet and its central body is provided by apparent gravitational attraction between them. [Real gravitational effort, pushing these bodies towards each other is considered here as apparent attraction between them]. Direction of this attraction, at datum points in the orbit, is perpendicular to bodies' linear motion. Action of the central force depends on the total magnitude of (radial) matter field- distortions, it is able to invest into the body's matter field. Magnitude of (perpendicular) matter field-distortions, a body is able to store is governed by its absolute linear speed. Action of apparent gravitational attraction is instantaneous and continuous. As long as participating bodies occupy common planes, apparent gravitational attraction on the bodies continues to invest distortions in their matter fields.

A planet obtains its constant radial speed, in the direction of the central body, by the time whole of its body crosses rear point on the central body and loses its radial speed in this radial direction during the time it crosses the foremost point of the central body. In the mean time, work introduced to create planetary body's radial motion is utilized to curve its linear path. Radial motion of a planet, moving in an orbital path, (for most of the time) is towards the central body. Real orbital path of a planet alternates on either side of its median path. Hence, its radial motion towards the central body is perpendicular to its linear motion only at inner and outer datum points. Following calculations corresponds to datum points. Magnitudes of spin actions at other points in the orbital paths vary according to the relative locations of central and planetary bodies.

By equation (7/11), remaining (radial) work, acting about the center of mass of a planetary body and producing its spin motion;

$$W_s = \frac{11MGmr}{16D^2V} - \frac{5MGmr}{16D^2V} = \frac{3MGmr}{8D^2V}$$

Radial work, $5MGmr \div 8D^2V$, acts to produce radial motion of the planet and the remaining work, $3MGmr \div 8D^2V$, acts acts to spin the body. Direction of radial motion in relation to body's linear motion varies throughout the orbital path. Hence, the magnitude and direction of the torque produced by the work will also depend on the relative position of a planet in its orbital path. However, with respect to perpendicular axis of the planet, direction of torque reverses every half of its apparent orbit. For whole period of each half-orbit, direction of torque remains steady and accelerates body's spin motion.

Since a planet is made up of composite materials and it is a large body, it cannot attain constant spin speed appropriate to this work, instantaneously. Instead, the work is held within the body as compressive energy, which is gradually converted to rotational kinetic energy. Rate of this conversion is body's spin acceleration and depends on the size and compressibility of the body.

Since the direction of central force, applied on the planet is towards the central body, direction of work, producing the spin motion, is towards the central body and it is applied on the rear hemisphere of the planetary body. As the planet develops spin motion, this work is distributed throughout the planet in various directions. Reduction in magnitude of original work is overcome by the continuous replenishment by the central force. Thus, irrespective of development of spin motion, spin-producing work in the body's matter field is augmented at a rate related only to the central force and its relative direction to body's linear motion.

7.5.1. Spin due to central force:

Part of the (radial) work, introduced into a planetary body's matter field, by the central force, creates planet's orbital motion. Leftover part of the (radial) work, after that is utilized for orbital motion, creates planet's spin motion.

Leftover radial work from the rear hemisphere, by equation (7/11);

$$W_s = \frac{3MGmr}{8D^2V}$$

Considering the action in any half-orbit, introduction of work is continuous at the rate given by the above equation (at the datum point). Magnitude of the work varies and depends on the location of the body in its orbital path. This work acts to accelerate the spin motion of the planet. Although the work is utilized according to its distribution throughout the body, to produce spin motion, continuous action by the central force prevents any depletion (other than by change in the relative direction) in its magnitude, in radial direction towards the central body.

As a planetary body moves along its orbital path about a central body, for every completed (small) orbital cycle (consisting of one complete apparent orbit), it is accelerating towards the median path in the direction of galactic center for half the time (shown in figure 7P, position A). Torque on the body (by part of the central force) during this period is in clockwise direction as shown in the figure by thick arrow. This is in the angular direction opposite to planet's (anti-clockwise, with respect to the central body) orbital motion. Planetary body develops clockwise (in negative angular direction) spin motion during this half of its orbital path, when it is moving from point E on median path through outer datum point to point D on median path, along the path EAD (as shown in the figure).

Similarly, for the other half of the orbital path, the planetary

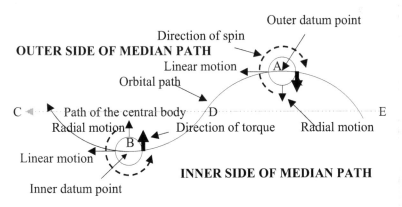

Figure 7P

body is accelerating towards the median path in a direction away from the galactic center (shown in figure 7P, position B). Torque on the body during this period is in anti-clockwise direction as shown in the figure by thick arrow. This is same as planet's (clockwise, with respect to the central body) orbital motion. Planetary body develops anti-clockwise (in positive angular direction) spin motion during this half of its orbital path, when it is moving from point D on median path through the inner datum point to point C on median path, along DBC. Spin motions, produced when the planetary body is on the outer side and on the inner side of median path are in opposite angular directions. They tend to compensate (neutralize) each other. Ultimately, over a long period, the planet will develop an overall resultant spin acceleration and spin speed in the direction of higher spin motion produced by the body, if any. Thus, depending on the parameters of the planetary body, it is possible for a planetary body to achieve spin motion in either direction or none at all.

By equation (7/11);

$$\text{Rate of increase in spin-work, } W_s = \pm \frac{3MGmr}{8D^2V}$$

As the spin-work increases in magnitude continuously, it is equal to a torque on the body. Let the spin acceleration be γ rad/sec^2. Spin speed of the body increases by γ radian in unit time. Angular displacement of the body in unit time due to spin-work (introduced in unit time) = $\gamma/2$ radian.

Total spin-work introduced in unit time = Torque × angular displacement in unit time

$$= \pm \frac{3MGmr}{8D^2V} \times \frac{\gamma}{2} \qquad (7/24)$$

$$\text{Kinetic energy due to spin} = \frac{\text{Moment of inertia} \times (\text{angular speed})^2}{2}$$

$$\begin{bmatrix}\text{Change in spin (Kinetic)} \\ \text{energy in unit time}\end{bmatrix} = \frac{I \times (\text{change in angular speed in unit time})^2}{2}$$

$$= \frac{I\gamma^2}{2} \qquad (7/25)$$

Equating the work as given by equation (7/24) to kinetic energy due to spin as given by equation (7/25),

$$\frac{I\gamma^2}{2} = \pm\frac{3MGmr}{8D^2V} \times \frac{\gamma}{2},$$

$$I\gamma = \pm\frac{3MGmr}{8D^2V}$$

Putting the value of $I = \frac{2mr^2}{5}$;

$$\frac{2mr^2}{5}\gamma = \pm\frac{3MGmr}{8D^2V}$$

$$\gamma = \pm\frac{15MG}{16D^2Vr} \qquad (7/26)$$

Considering the action in unit time; Let γ be angular acceleration of the body's spin motion (due to radial work) at the datum points in the apparent orbit. Direction of radial work reverses every time the body traverses median path.

In equation (7/26), factors V and D are variables, varying continuously as a sine function of relative angular position of the planetary body with respect to the central body. For approximate calculations, their average values may be used. Linear component of relative velocity varies in proportion to the sine of planet's relative angular displacement with respect to central body, taking a point in the orbit on the median path as reference. Mean value of relative linear motion is equal to $2v/\pi$ times that of the highest value of relative velocity, v, at datum point. To get the mean linear speed of the planet on the outer half-orbit, V_{out}, mean value of relative speed at outer datum point is added to value of linear motion of the body, when it is on the median path (which is the same as the linear speed of the central body, V_c). To get the mean linear speed of the planetary body, V_{in}, on the inner half-orbit, mean value of relative speed at inner datum point is subtracted from the value of linear motion of the body, when it is on the median path (which is the same as the linear speed of central body, V_c).

Mean linear speed of planet on the outer half-orbit,

$$V_{out} = V_c + \frac{2v_{out}}{\pi} \qquad (7/27)$$

Mean linear speed of planet on the inner half-orbit,

$$V_{in} = V_{aphe} + \frac{2v_{in}}{\pi} \qquad (7/28)$$

Similarly, we may calculate the mean distances between the bodies by using the multiplicand factor $2/\pi$. If deflections of perihelion/aphelion from datum points are very small, mean distances may be found by using D_{peri} and D_{aphe} instead of distances between bodies at datum points.

Mean distance between bodies, on outer side of median path,

$$D_{out} = D_{peri} + \frac{2(D_m - D_{peri})}{\pi} \qquad (7/29)$$

Mean distance between bodies, on inner side of median path,

$$D_{in} = D_m + \frac{2(D_m - D_{aphe})}{\pi} \qquad (7/30)$$

Substituting these values of D_{out} and D_{in} in lieu of D in equation (7/24), we get approximate average spin acceleration γ_{out} of the planet on the outer half-orbit and approximate average spin acceleration γ_{in} of the planet on the inner half-orbit.

$$\gamma_{out} = -\frac{15MG}{16 D_{peri}^2 V_{out} r} \qquad (7/31)$$

$$\gamma_{in} = +\frac{15MG}{16D_{aphe}{}^2 V_{in} r} \qquad (7/32)$$

Since the spin accelerations in outer half-orbit and inner half-orbit are in opposite directions, angular displacements produced by them during each half-orbit are in opposite directions. If it is clockwise, when the body is on the outer side of median orbital path, it will be anti-clockwise, when the planet is inside the median orbital path. (To be exact; the median orbital path is along the line joining the mid-points between outer and inner datum points, in the orbital path. If we ignore the lateral motion of the central body, the planetary body crosses the path of central body at the mid-points of the orbit, points E, D and C as shown in figure 7P. At every mid-point in the orbital path, radial motion due to central force and linear motion due to inertia of a planet are co-linear).

When considering the spin motion, over a full (apparent) orbit, the body may have an overall resultant angular acceleration (angular displacement) in any one direction or (in the rare cases, when spin motions on either sides are equal to each other) it may not have a resultant spin motion at all.

Spin acceleration in the half-orbit, outside the median path of the planet, is in opposite angular direction to planet's orbital motion. Spin acceleration in the half-orbit, inside the median path of the planet, is in the same angular direction as the planet's orbital motion. They are in opposite angular directions. Hence, overall resultant spin displacement of the planet depends on the time the planet spends in either of the half-orbits and the magnitudes of angular accelerations in either of the half-orbit. In the outer half-orbit, a planet travels at a greater linear speed but the distance traveled is much greater. In the inner half-orbit, the distance traveled and body's linear speed are lower. Linear acceleration / deceleration of a planet are (approximately) uniform. Hence, the time spent by the planet on either side of its median path is proportional to the highest / lowest relative velocity (at datum points) achieved by the body. Let T be the orbital time period of the planet and let t_{out} and t_{in} be the time spend by the body on the outer side and inner side of median path, respectively. v_{in} is the relative velocity at inner datum point and v_{out} is its relative velocity at outer datum point.

Time spent by the planet on the outer side of median path,

$$t_{out} = \frac{V_{out}}{V_{out} + V_{in}} \times T \qquad (7/33)$$

Time spent by the planet on the inner side of median path,

$$t_{in} = \frac{V_{in}}{V_{out} + V_{in}} \times T \qquad (7/34)$$

Angular displacement, due to spin acceleration, in the outer half-orbit,

$$\theta_{out} = -\gamma_{out} \times \frac{t_{out}^2}{2} \qquad (7/35)$$

Angular displacement, due to spin acceleration, in the inner half-orbit,

$$\theta_{in} = \gamma_{in} \times \frac{t_{in}^2}{2} \qquad (7/36)$$

Medium sized planets may have thin crust in solid state. Otherwise, all large bodies are fluid in structure. Alternating spin motions of these bodies are effectively dampened by churning of fluid parts in these bodies. For half a planetary year, the spin acceleration is in one direction and for the next half of the planetary year, the spin acceleration is in the opposite direction. Only an overall resultant spin motion is noticed over a long period of time. It may be averaged for a full orbital period to give overall average spin acceleration, Γ, of the body.

$$\Gamma = 2(\pm \theta_{in} \mp \theta_{out}) \div T^2 \qquad (7/37)$$

Direction of this spin motion depends on the orbital characteristics of the body. Hence, it is quite natural for planets to spin in either direction in its apparent orbital plane or not to spin at all.

Usually, all planets have eccentric apparent orbits in the same direction and hence they tend to spin in the same direction as their orbital motion. As the eccentricity of apparent orbit reduces or changes direction, differences in linear speeds of the planet, when on either side of median path and the differences in the duration spend on either side of median path become less. Difference between clockwise and anti-clockwise spin motions reduces. In case, $v_2 \gamma_{in} = v_1 \gamma_{out}$, spin displacement in either direction are equal and the planet will have no resultant spin motion at all. At very low eccentricity of the apparent orbit, clockwise (anti-orbital) spin motion produced on the outer side of median path may become more than the anti-clockwise (pro-orbital) spin motion produced on the inner side of median path. Such bodies angularly accelerate in opposite direction to their orbiting direction. They spin in counter direction. Should a planet have one or more satellites, these satellite bodies also exert central forces on the planet to produce spin motions. Average spin motion of the planet is the resultant of all spin motions produced about it.

Spin acceleration of a planetary body is caused by the perturbations in its path. There are no actions (except those due to external interference) available to oppose or modify this phenomenon. Hence, a planetary body with overall resultant spin acceleration (in any direction) will continue to accelerate in its spin motion indefinitely. Energy for this motion is directly derived from the kinetic energy of the body due to the central force. Therefore, development of spin motion does not affect body's linear motion in its orbital path.

Perpetual spin acceleration of a planet will ultimately lead towards its disintegration under tidal and centrifugal stresses. This phenomenon prevents eternalness of planetary systems in nature and contributes towards re-cycling of 3D matter to reduce universal entropy.

Exemptions to the above explanations may be observed in the following cases. In case a planetary body already had a spin motion (before entering into its orbit) in opposite direction to that is created by its planetary motion, the body will first slow down its spin motion to a stop and then reverse the direction as directed by the torque acting on it. Spin motion of the planetary body, during the transition stage, may not correspond to the parameters of the body and its orbital parameters. Spin motions of the planetary body, in any other directions, will also be modified in due course of time.

Bodies, with spin motions outside their orbital planes are relatively new additions to the planetary systems. (Disregarding the initial conditions), time spend by a planet in a planetary system may be estimated from its current spin speed and orbital and body parameters.

7.5.2. Unequal spin motion of a planetary body:

Equation (7/31) and equation (7/32) give overall spin accelerations of a planet. Work, acting to produce these accelerations, is concentrated towards the equatorial regions (at the rear) of the planetary body. Hence, these parts of the planets tend to move first and carry the rest of the body along with it, gradually, due to integrity of the body. How fast the rest of the body attains the same spin speed as the equatorial region depends on the rigidity of the body.

Usually, all planets are spherical in shapes. Actions, causing the planet's spin motion, are concentrated at its equatorial region. Since parts of a solid body cannot have relative motion between them, action of work at one part of the body is transmitted to other parts very fast and whole of the body tend to move at the same angular acceleration. If a planet (or the central body) is in fluid state or it has large fluid outer cover, its equatorial region, where (spin) work is concentrated will move first and rest of the body will gradually pick up the motion. More viscous a planet is, less delay in the follow up. There will always be a delay in the motion of the rest of the body. In other words, equatorial region of a fluid planet will lead other parts in spin motion.

All very large bodies are in fluid state or have fluid outer surfaces. Therefore, the equatorial regions of all large bodies in a multi-body system, have higher spin speed compared to spin speeds at their polar regions or inner parts. This phenomenon is clearly noticed in the cases of sun and larger planets of solar system.

7.5.3. Apparent spin motion:

'Absolute spin motion' of a planetary body is with respect to the 2D energy fields in space. To create this spin motion, it requires external effort and it is produced by the action of apparent gravitational attraction on the body. Work invested by the external effort, to produces the spin motions of planetary bodies, change their states of motion. This work is stored in the body in the form of kinetic energy due to spin motion. 'Apparent spin motion of a planetary body' is with respect to any reference the observer assigns. This requires no external effort on the planetary body. It is only apparent and may change with a change of reference. Usually, apparent spin motion is what we observe in every day life. Combining the apparent spin motion of a planetary body with its absolute spin motion, often results in false notions.

Central body develops a spin about one of its diameters perpendicular to the plane of the orbits of all its planetary bodies. Spin speed and direction of a central body are the resultant of spin speeds produced on it by all its orbiting planetary bodies. A stable central body cannot accommodate planets or satellites in different orbital planes. Hence, an ideal planetary system about a central body has only one orbital plane and the central body and all planets and satellites in the system have their spin axes perpendicular to this plane.

As a planet apparently revolves around the central body, there is no force on it to maintain relative positions between a surface point on the planet and a surface point on the central body. As the planet appears to move around the central body, the planet makes one apparent rotation every revolution in its apparent orbit. This 'apparent spin motion' of planet requires no external effort or energy. Body appears to spin about one of its diameters, perpendicular to the plane of the apparent orbit, once every completed apparent orbit (one planetary year). Direction of this apparent spin motion (in case of bodies, which spin in opposite direction to their orbital motion) is such that a point on the surface of the planet, away from the central body, will appear to move in the direction opposite to the orbital motion of the planet. In case of bodies, which spin in the direction of their orbital motion, the apparent spin is in opposite direction to its real spin motion. Should these spin motions be equal to each other, the planet appears to have no spin motion at all. That is: same face of the planet remains permanently towards the central body. However, over extended periods, planet's own spin motion will supersede the apparent spin motion. In case of those bodies, which spin in opposite direction to their orbital motion, the apparent spin motion is in the same direction as its real spin motion.

Figure 7Q shows relative positions of a spinning planet and its central body, one planetary-day apart. Line PC and P_1C_1 join the centers of these bodies. 'H' is a reference point on the surface of the planet, facing the central body. G shows its position at the same time on the next day. A planetary-day is the time elapsed between subsequent instants when reference point 'H' faces the central body. For this, reference point 'H' has to be on line P_1C_1 after a lapse of one planetary-day from the time it was on line PC. Let the planet be spinning at a constant speed in the direction,

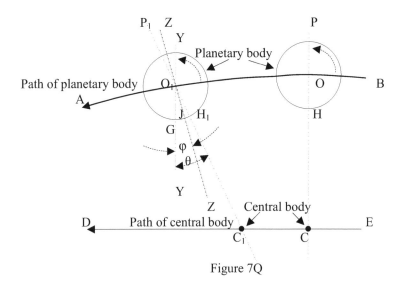

Figure 7Q

shown by the dashed curved arrows. By the time, the central body moves from C to C_1, the planet moves from O to O_1. In the mean time, the planet would have turned through 2π radians. This takes the reference point, H, to position G in the figure. Although the planet has rotated through one full turn, in order to complete a planetary-day, the planet has to have an additional angular displacement from G to H_1. θ, is the additional angular displacement required every day to maintain constant length of planetary day. This additional displacement is apparent because the body only appears to have lost same magnitude of turning motion in a planetary-day, relative to the central body. No real angular displacement or motion of the planetary body takes place. Loss of turning motion is caused by equating a planetary-day and a rotation-day of the planet. This negative angular displacement is caused by the curvature of planet's orbital path rather than its spin motion. A planet has similar apparent spin motions due its orbital motion about the central body and due to its orbital motion about the galactic centre. If they are in the same direction, they augment to each other. if they are in opposite directions, they nullify each other.

Since a planetary-day is related to the apparent spin motion of the planet with respect to the central body, the above discrepancies produce constant difference from the rotation time of the planet and vary the length of planetary-day from that of its rotation-day. In order to compensate for this, the planet has to have an additional angular displacement θ, every day. A planetary-day needs an additional (apparent) spin movement of the planet added to its rotation-day. If the compensation provided for the additional angular displacement of the body, in a planetary-day is equal to the apparent angular displacement, θ, length of planetary-days remain constant, as long as the compensation balances the discrepancy.

If the additional angular displacement added for a planetary day, is more than the apparent angular displacements (it over-compensates the discrepancy, θ), the reference point will cross the point H_1 before a planetary-day is completed. Length of a planetary-day appears to have shortened and the planet appears to be accelerating in its spin motion.

If the additional angular displacement added for a planetary-day is less than the apparent angular displacements (it under-compensates the discrepancy, θ), the reference point will not quite reach the point H_1 on completion of a planetary-day. Length of a planetary-day appears to have increased and the planet appears to be decelerating in its spin motion. We observe the earth in this apparent condition.

Since a planetary body constantly accelerates in its spin motion, compensation provided to equalize rotational-day and planetary-day, requires constant modification.

7.5.4. Anomalies:

Once a planet has settled into its stable orbit about a central body, its spin motion is automatically developed, maintained and accelerated. Variations in its parameters introduce modification of inertial actions and thereby modify the spin speeds of both central and planetary bodies. If these bodies are massive and spinning at relatively high speeds, they will also have the property of gyroscopic precession. An external effort (by collision with another body), acting on them will invoke precessional response. This causes wobbling of axes of bodies in multi-body systems. In extreme cases, this wobbling may reach up to 90° to the line of orbit (as in the case of Uranus). Wobbling, introduced by an external effort, can be removed from the body only by applying an equal and opposite external effort on the body. If the magnitude of wobbling is high, magnitude and directions of spin motions to the planet may be greatly altered.

Should there be more than one body, orbiting about a central body in nearby orbits in the same direction and these bodies are near enough, they are under considerable mutual apparent attraction due to gravitation. They may approach each other simultaneously, as they move in their own orbital paths. Since their momentum towards each other is relatively small, these bodies may gradually collide into each other and merge. If these bodies were in orbit about the central body for a length of time, they would have gained and adjusted their spin speeds according to their orbiting speed and other effective parameters. Collision between two spinning bodies (with similar spin direction) is bound to reduce or nullify spin motion of the both bodies.

Due to opposite directions of motion of touching surfaces, most parts of these planetary bodies will be torn off their surfaces. Fragments flying away from the site of such collision and moving in the same angular direction as other planets of the union at the right linear speed could form a dust belt about the central body. All parts of the dust belt will move in a common orbital path.

If a planetary body undergoes more collisions of similar nature, it is possible for the resultant (remaining) body to spin at low speed or even spin in the opposite direction. In due course of time, this will be rectified by gradual inertial actions.

7.5.5. Variations in a terrestrial day:

Irrespective of its constant spin acceleration, the earth is observed to decelerate in its spin motion. This is an apparent phenomenon due to inability of additional angular displacement (created by earth's spin acceleration together with compensations provided) to fully compensate earth's apparent spin motion. This shortage, being extremely small, is not normally noticed. However, over extended periods of time, earth's solar day appears to expand. Earth takes more time to complete one solar day. This is misinterpreted as slowing down of earth's spin motion.

From the above explanation, it is seen that the lengthening or shortening of a planetary-day or the apparent deceleration or acceleration of spin speed has no relation to the variations in the real spin speed of the planet. Spin motion of a planet can only accelerate (other than in cases, where the original spin motion of the planet is in opposite direction to its natural spin motion). Magnitude of spin motion depends on the curvature of its path and effective diameter of the body. Curvature of orbital path, at different points along the orbit, varies. This is bound to vary the spin acceleration of the planet on periodical basis.

Similar argument is applicable to motion of planetary system's central body also. Central body develops spin motion about one of its diameter perpendicular to the plane of its own orbit about the galactic centre. Additionally, the central body acts as an orbiting body about each of its planets. This creates additional spin motions. If there are more than one planet, spin speed of a central body will be the resultant of spin-speeds produced by all its planetary bodies and spin speed due to its own orbital motion. A stable central body cannot accommodate planets in orbital planes differing from its own. If a planet pairs up with central body in a different plane, central force between them will have an additional component, acting on the planet, to modify its orbiting parameters. Gradually they will fall in line. Hence, in a planetary system there is only one stable orbital and spinning plane. Central body and all planets in the system have their spin axes perpendicular to this plane. Uneven distribution of mass in bodies of collective system may introduce smaller individual movements (periodical or otherwise) of these bodies within small limits. External efforts on the bodies may introduce wobbling / tilting of their axes.

7.6. Tides:

The term 'tide' generally means the distortion in the shape of a linearly moving-spinning body. On earth, it is also considered as periodic rise and fall in the sea level around earth's sphere. We may concentrate our attention to the tides produced on earth by external efforts.

Matter content of a planetary body, orbiting about a central body, is tied up to the matter content of central body by the central force. Direction of the central force, due to apparent gravitational attraction, is towards the central body and there is no other external effort acting on the planetary body. Every particle in the planetary body is attached to central body by the central force. Hence, all particles in the planetary body separately and together are at equilibrium. There can be no relative motion between them due to the central force. Hence, the assumption that the tides on earth are caused by apparent gravitational attraction towards the sun (and the moon) is not valid. It is also illogical to assume that equal tide on the opposite sides on the surface of earth is produced to balance the shift in mass, by the same external efforts.

In a free and stable two-body system, there is only one external effort on each of the bodies and acting in between them - the central force. Central force between the central body and the planetary body is the apparent attraction between the bodies due to gravitation. It is the result of separate gravitational push forces on both the bodies, towards the other. There is no other external effort on the earth, which is moving in an apparent circular orbit (we may ignore the eccentricity of the apparent orbital path, for explanation on tide) about the sun or on the moon moving in an apparent circular orbit about the earth. [Centrifugal force is an imaginary force fashioned by the inertial effects of a body, moving in a circular path].

All inertial actions on a body is the product of surrounding 2D energy fields. Another body in the neighborhood does not directly affect the body, by its inertia. All it can do is to affect the surrounding 2D energy field by its presence. Modification in the 2D energy field may cause variation in its actions on the first body. This is usually understood as a direct interaction between the two bodies.

Apparent gravitational attraction between a central body and its orbiting body produces a central force between them. The same effort of apparent gravitational attraction cannot produce partial or additional external efforts of attraction between these bodies (for the sake of the tides). Therefore, all matter particles in the body of earth remain in place relative to each other and no phenomenon like tide can take place, unless there is some other phenomenon, acting on the body-matter of the earth, without affecting the central force or its actions on the body. Therefore, this localized phenomenon, which produces tides, is not related to the apparent gravitational attraction between the sun/moon and the earth. Phenomenon of tide affects only an orbiting body that spins. If the cause of tide on an orbiting body with spin motion is the apparent gravitational attraction towards another orbiting body with spin motion, phenomenon of tide is applicable to that body also. The central body, being an orbiting body with respect to the planetary body, also is affected by the phenomenon of tide. That is to say that the phenomenon of tide is the product of parameters of a spinning-orbiting body, alone. These parameters may be modified by the presence of other bodies in the vicinity.

Let an external linear effort act on a spinning body (with no linear motion), along a radial line of its matter field, where the arms of 2D energy field latticework squares are symmetrical about the direction of external effort. Since the external effort produces no angular component, it cannot affect body-particles' tangential motion. Body has to develop pure linear motion in addition to its original rotary motion. All matter particles of a spinning body move in curved paths about the body's centre of rotation. On one side of the centre of rotation, matter particles of the rotating body have newly introduced linear motion assisting their tangential motion in the curved paths. On the other side, matter particles of the rotating body have newly introduced linear motion opposing their tangential motion in the curved paths. Centre of rotation of the spinning body shifts without affecting the angular speed of the body. As the additional linear motion increases, centre of rotation of the body can move even outside the body-material. Angular and linear motions of the macro body and corresponding distortions in its matter field remain independent irrespective of changes in any of them. Newly introduced linear motion cannot affect body's rotary motion. In order to satisfy this requirement, inertial actions in the body's matter field control the curvatures of the body-particles' path (in space), by appropriate changes in their directions of motion. Changes in the curvature of paths, with the tangential speed remaining constant, requires shift in the centers of their curved paths. During linear accelerating stage of a rotating body, its radii in various planes, perpendicular to its spin axis, vary all around. Circular cross-sections of the spinning body (perpendicular to spin axis) attains elliptical shape. The body bulges outwards along the line of external effort's action. Body's shape will revert to original state as the body attains a steady state of constant linear and spin motions.

Figure 7R shows a circle in dotted line, which represents the equatorial plane of a rotating macro body. A, B, C, D, E, F, G and H represent resultant distortions in few of the matter field squares in different locations in the body. Although, they are represented in the shape of latticework squares, they represent the directions of reactions developed in the latticework squares of macro body's matter field. (All additional works in a macro body have to be conveyed only through the arms of 2D energy field latticework squares). Resultant distortions, shown in the figure, represent the spin-part of resultant additional distortions in the matter field, rather than the actual matter field squares in the macro body. Curved arrows in dotted line show the direction of spin motion – anticlockwise – of the macro body. Let the spin speed of the rotating body is equal to +ω units/sec. Small arrows, in vertical direction, show parts of an external effort applied, simultaneously and evenly, on the rotating body. The external effort is applied equally on every matter field latticework squares. All junction points of matter field latticework experience equal resultant actions in the direction of the applied effort. Although the actions of the effort is to introduce identical linear distortions in every matter field latticework squares

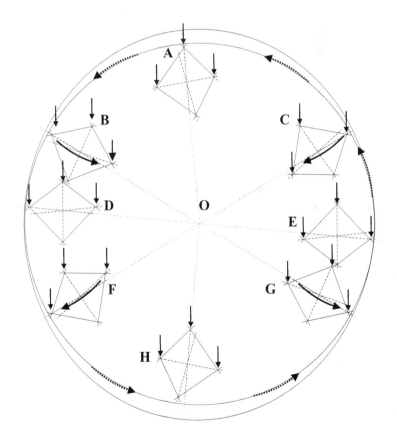

Figure 7R

of the body, orientation of the matter field squares at different locations, due to resultant distortions in them, cause slight differences in the magnitudes of action at different matter field latticework squares.

At locations, A, D, E and H, arms of matter field latticework squares (towards the source of external effort) are symmetrical to the external effort. Action at all junction points and the distances between junction points, in perpendicular direction to the external effort, are equal. Distortions introduced by the external effort are purely linear and the work introduced is used solely for linear motion of the rotating macro body in the direction of the external effort.

At locations, C and F, orientation of a matter field square (distortion) is deflected anticlockwise. Distances between middle junction point and junction points on either side are different. Junction point to the right is farther than the junction to the left. Although the magnitudes of vertical actions are same, difference in the distances to the junction points on the sides of the latticework square, produce a turning movement of the matter field square in clockwise direction, $-\alpha$, as shown by curved arrow in bold line. Direction of this deflection is in opposition to the matter field-distortions, producing body's spin motion. Hence, it tends to reduce the spin speed of the body-particles in these locations.

At locations, B and G, orientation of the matter field latticework square (distortion) is deflected clockwise. Distances between middle junction point and junction points on either side are different. Junction point to the left is farther than the junction to the right. Although the magnitudes of vertical actions are same, difference in the distances to the junction points on the sides, produce a turning movement of the matter field square in anti-clockwise direction, $+\alpha$, as shown by curved arrow in bold line. Direction of this deflection is in the same direction of the matter field distortions, producing body's spin motion. It tends to enhance the spin speed of the body in these locations. The gray dashed lines PO and RO in the figure 7R show median lines, of the arms of the matter field squares.

Turning movement of matter field distortions, explained above also causes angular displacement of body-particles. Therefore, effect of the turning motion of matter field squares is also measured in terms angular displacement of matter particles. However, its action on the body is without affecting tangential speeds of body-particles, which produce body's spin motion. For this, the curvatures of the paths of matter particles of the body are modified without altering their tangential speeds. A matter particle in the body moves outward from O (the centre of rotation in figure 7R), during its travel from E to A. It moves inward towards O, during its travel from A to D. It moves outward from O, during its travel from D to H. And it moves inward towards O, during its travel from H to E. Path of the body-particle is shown in bold ellipse in the figure 7R.

Changes in the paths of the body's matter particles alter body's shape during the action of the external effort on the body. (Circular) equatorial plane and all (circular) planes parallel to it assume elliptical shape. Change in the shape of the body lasts only during rotating body's linear accelerating stage. Once the external effort is terminated and accelerating stage is over, the body will reach a steady state of combined motions of spin and linear motions and it will revert to its original spherical shape. During linear acceleration stage of a spinning body, it elongates along the direction (forward and rearward) of the external effort. This phenomenon causes 'tides' on spinning planetary bodies.

Phenomenon of 'tide', on a linearly moving spinning body, is produced by its response to an external effort. As long as an external effort is acting on the body, tides on one body are not related to any other body. Absolute motions of a body (with respect to 2D energy fields) produce the tide. Since we have no absolute reference (until the universal medium is recognized), it is almost impossible to determine absolute speeds of a body. Directions of spin axis and orbiting plane of a planetary body, magnitude of water body on its surface, latitude of surface point considered, relative position of planetary body in its orbital path with respect to its central body, any independent motions, etc. also affect the magnitudes and relative positions of the tides on a body. All these factors, together with the number of independent external efforts on a planetary body determine the relative position and magnitude

Chapter 7. GENERAL

of resultant tides at any point on the surface of a planetary body. [Refer to parent book for explanation on linear motion of a rotating body].

Phenomenon of tide takes place only during accelerating stages of a macro body, which already has a steady state of motion. Spin acceleration of a linearly moving body or linear acceleration of a spinning body, change body's shape to produce tide. Since the tide is not related to body's steady state of motions, each of the external efforts acting on the body produces its own sets of tides on the body. According to current theories, two external linear efforts acting on a body produce only one resultant motion of the body. Yet, in case of terrestrial tides, we come across two separate external efforts acting on a rotating body and they are able to produce two sets of tides at different points on the body. This clearly contradicts any explanation of tide, related to the displacement of parts of the body in the direction of external effort. Further, this concept (according to which the phenomenon is described here) does not recognize pull forces that acts through empty space to pull at parts of a rotating body and thereby to create tides.

No free body in space can remain static. They move, mostly in curved paths. Let us consider a spherical spinning body in space moving in a linear path. A particle on the equator of the body traces a circular path about body's spin axis. Simultaneously, the particle is carried with the body in its linear motion. Circles in figure 7S show representations of an equatorial plane of a spinning body, moving in linear direction from A' to A. The body moves linearly in the direction of linear arrows while spinning in the direction of curved arrows. A 2D matter particle (plane section of a 3D matter particle) at A, on the equatorial surface of the body, is

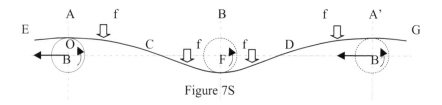

Figure 7S

carried along the curved path GA'DFCAE in space. This particle has a constant angular speed, ω, about the spin axis of the body. As shown, the angular speed is anticlockwise (+ω).

Let a constant external effort, f, shown by the block arrows, act on the body continuously. As long as they are in the same plane, relative direction of external effort to the direction of linear motion of the body does not affect the magnitude of tides formed. Directions of tides depend on the direction of the external effort, f. Tides may be formed in the direction of linear motion of the body, also, provided the external effort is in that direction. All 3D matter particles in the body are affected identically. Depending on the relative direction of external effort, they are linearly accelerated or decelerated in the direction of external effort.

Work (distortions), introduced by the external effort, in body's matter field has to accommodate itself within the additional work already existing in the body and producing body's linear and rotary motions. Newly introduced additional distortions modify the existing additional distortions during the accelerating stage. Since the external effort is of constant magnitude and continuous, modification is at constant rate. Part of the linear distortions due to external effort, f, (that attained stability) causes the particle to move or change its steady state of motion (at constant speed) in the direction of external effort. Changes in the matter field-distortions during transition-stage accelerate the 3D matter particles. Component of linear acceleration (about the centre of rotation), aiding/opposing the tangential motion, acts to modify the angular displacement of the 3D matter particles along the curved path. Direction of angular acceleration depends on the relative position of a 3D matter particle about the centre of curved path. Resultant angular speed of the 3D matter particle along the curved path is modified accordingly, without affecting its tangential speed.

With reference to the figure 7S; consider a 2D matter particle (matter content in a plane of a 3D matter particle) situated at a point, A', on the surface of linearly moving spinning body. During the body's motion along the central line shown in the figure (motion of a planet in its orbital path), the 2D matter particle is carried along curved line GA'DFCAE;

- From A' to D, the 2D matter particle is on the left side of the centre of rotation and hence undergoes anticlockwise angular acceleration (+ α). Total angular speed of the particle in its curved path is increased. The 2D matter particle tends to move towards the centre of rotation of the spinning body.

- From D to F, the 2D matter particle is on the right side of the centre or rotation and hence causes clockwise angular acceleration (- α). Total angular speed of the 2D matter particle in its curved path is reduced. The 2D matter particle tends to move away from the center of rotation of the spinning body.

- From F to C, the 2D matter particle is on the left side of the centre of rotation and hence causes anticlockwise angular acceleration (+ α). Total angular speed of the 2D matter particle in its curved path is increased. The 2D matter particle tends to move towards the center of rotation of the spinning body.

- From C to A, the 2D matter particle is on the right side of the centre of rotation and hence causes clockwise angular acceleration (- α). Total angular speed of the 2D matter particle in its curved path is reduced. The 2D matter particle tends to move away from the center of rotation of the spinning body.

Matter contents in many 2D planes constitute a 3D matter particle. Because of motions of matter contents in various planes (2D matter particles), every 3D matter particle in the body travels along similar paths. As a result, the body bulges outwards in both directions along the line of action of the external effort. Spherical body's equatorial plane changes its shape from a circle to an ellipse.

Figure 7T shows the transformation of the equatorial plane of a linearly moving (spherical) spinning body, ACBDA, under the action of an external effort, f, shown by the thick arrow, evenly applied on whole of the body. Curved arrow shows the direction of spin of the body.

Let; The spin speed of the body = ω rad/sec,

Rotational time period of the body = T sec,

Radius of the body = r meters,

Tangential speed of a 2D matter particle = v m/sec,

Magnitude of external effort = f kg-m/sec^2,

Mass of a 2D particle of the body situated at the surface on equatorial plane of the spherical body at D = one unit.

Linear acceleration of 2D matter particles in the body due to external effort = a m/sec^2.

Displacement of 2D matter particle in unit time = a/2 m (displacement = at^2/2)

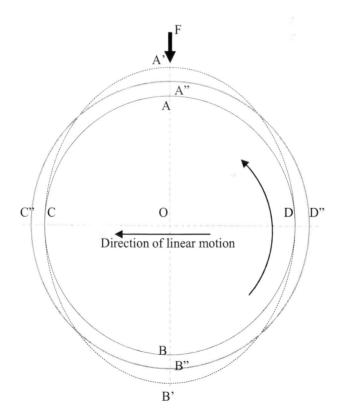

Figure 7T

Additional angular speed of the 2D matter particle in its curved path at D,

$$\omega_a = a/2r \text{ rad/sec} \qquad (7/37)$$

Relative direction of external effort with respect to the path of the 2D matter particle changes as the 2D matter particle moves along the curved path. Angular acceleration of the 2D matter particle, produced by the external effort, varies relative to direction of action of the effort. Magnitude of angular acceleration is highest when the particle is at D or at C and it is lowest when the 2D matter particle is at A or at B, in the figure 7T. Magnitude of angular acceleration with respect to spin axis varies in proportion to the cosine of the angular displacement of the 2D matter particle from D. Therefore, the mean magnitude of angular acceleration in any quadrant is equal to $\pi/2$ times of the highest magnitude of angular acceleration produced at C or D.

Mean magnitude of additional angular speed,

$$\omega_{a\,mean} = \frac{a}{2r} \times \frac{2}{\pi} = \frac{a}{\pi r} \qquad (7/38)$$

Additional angular speed attained by the particle during its travel from D to A

$= \omega_{a\,mean} \times$ time to move through one quadrant

$$= \frac{a}{\pi r} \times \frac{T}{4} = \frac{aT}{4\pi \times r} \qquad (7/39)$$

- From D to A, additional angular speed is in opposition to the particle's original angular speed.
- From A to C, additional angular speed is in the direction of the particle's original angular speed.
- From C to B, additional angular speed is in opposition to the particle's original angular speed.
- From B to D, additional angular speed is in the direction of the particle's original angular speed.

Resultant angular speed of the particle at A and B

$$= \omega - \text{additional angular speed}$$

$$= \omega - \frac{aT}{4\pi \times r} \qquad (7/40)$$

Resultant angular speed of the particle at C and D

$$= \omega - \frac{aT}{4\pi \times r} + \frac{aT}{4\pi \times r} = \omega$$

Angular speeds of the particle are lowest at A and B.

$$\left(\begin{array}{c}\text{Radii of circular}\\ \text{path at A and B}\end{array}\right) = \frac{\text{Tangential speed}}{\text{Resultant angular speed}}$$

$$= \frac{v}{\omega - \dfrac{aT}{4\pi \times r}} = \frac{4\pi \times rv}{4\pi \times r\omega - aT} = \frac{rv}{v - \dfrac{aT}{4\pi}} \qquad (7/41)$$

$$\begin{bmatrix}\text{Increase in the radii of}\\ \text{curved path at A and B}\end{bmatrix} = \begin{bmatrix}\text{Radius of circular}\\ \text{path at A or B}\end{bmatrix} - \text{Original radius}$$

$$= \frac{rv}{v - \dfrac{aT}{4\pi}} - r = \frac{rv - rv + \dfrac{raT}{4\pi}}{v - \dfrac{aT}{4\pi}} = \frac{raT}{4\pi \times v - aT} \qquad (7/42)$$

Increase in the length of body's diameter in the direction of external effort varies in proportion to the magnitude of action of the external effort and to the tangential speed of matter particles (spin speed) of the macro body. Magnitude of external effort being constant, higher rotational speed tend to increase the magnitudes of tides. Higher rotational speed increases the tangential speed of matter particles and reduces their rotational period. As the denominator of equation (7/41) approaches zero, magnitudes of tides increase considerably and may affect the integrity of a spinning body.

In figure 7T, ACBDA in dotted line is the original shape of the equatorial plane of a spinning macro body, without linear motion. Elliptical path (in dashed line), DA'CB', shows the path of a surface particle on the spinning macro body. Changes in the circular paths of the constituent particles of the macro body (due to the action of linear external effort, f) change the shape macro body's equatorial plane. Similar actions take place in all planes perpendicular to the spin axis. Spinning macro body elongates along the direction of action of the external linear effort. Circle A"C"B"D"A" is the mean assumed shape of the macro body. A"A' and B"B' show the heights of high tides and CC" and DD" show the depth of low tides from the assumed mean equatorial surface of the spherical macro body.

From the above explanations, it can be seen that the changes in the macro body's matter field change the body's shape in correspondence with the external effort on the body. Magnitude of change depends on the magnitude of the external effort and the spin speed of the body. Change in shape of the macro body is simply due to the rearrangement of its matter field during acceleration period rather than any motion of the body or relative displacements of its constituent matter particles. Hence, any number of forces, acting on the macro body, will introduce as many sets of distortions in the shape of the body (under linear motion and/or spin motion) as there are external efforts. Even while the macro body is moving linearly, it is able to have tide effects simultaneously in many directions.

Tides are distortions induced by an external linear effort (force or torque) in the shape of a spinning body, moving or not moving in linear direction. As soon as an external effort commences acting on a spinning body, the body would acquire a linear motion. Sources of the external effort or consistency of rotating body is immaterial to development of tides. All natural efforts are of push nature. Tides are produced by linear efforts on rotating bodies or by rotating efforts on linearly moving bodies. Both types of motions, linear and rotary motions, are involved. Otherwise, the external linear effort simply produces linear acceleration of the body and external torque simply produces spin acceleration of the body.

If a spherical spinning body is under constant action of an external linear effort, perpendicular to its spin axis, cross sectional planes (circular planes perpendicular to the spin axis) of the spinning body will change to and maintain their elliptical shapes. Major axes of the ellipses will be in the direction of the external effort. This makes the rotating body bulge outwards in both directions, towards and away from the direction of the incoming external linear effort. Increase in the diameter of the rotating body (in cross sectional planes perpendicular to spin axis) due to the bulges in the direction of external linear effort creates the phenomenon of 'tide'.

If a spinning body is of uniform consistency, there is no relative displacement of body particles due to tide. Curved path of each particle, with respect to centre of rotation of the body, is modified to suite the present requirements. Body particles are not attracted towards or repelled away from the source of external effort. They are not displaced, with respect to their neighbors, in any direction to create the phenomenon of tide about the spherical body. Observed effect is a simple change in the shape of body's matter field with corresponding re-shaping of the paths of body's matter particles, in space. This appears as tide, about the spinning body. Similar action takes place also during angular acceleration due to the action of a torque on a body under linear motion. Once the accelerating stage is over, the body and its shape will settle down to its steady states in both linear and rotary motions. The tidal effects are no more present during body's constant state of motions.

After linear acceleration period, a rotating body settles down to a steady state of combined motion of linear and original spin motions. Due to linear speed, attained by the rotating body, matter field-distortions on one side of its centre of rotation increase their transfer speeds while matter field-distortions on the opposite side of its centre of rotation reduce their transfer speeds. Location of the matter field latticework square that produces no angular deflection (centre of rotation of the body) is shifted, from the centre of mass of the body, towards the side of the rotating body, where body-particles were moving in opposite direction to the newly introduced additional linear motion.

7.6.1. Terrestrial tides:

Earth is a spinning macro body, moving linearly along its orbital path about the sun. Central force between the sun and the earth guides the earth along its curved orbital path. The earth is also under another central force from its satellite - the orbiting moon. Since the rest mass of the moon is very small compared to that of earth, perturbation it can cause to earth's orbital path is very small. Instead, moon's orbital path conforms to perturbations produced by the central force between earth and the moon. However, central force between earth and the moon also acts as an external effort on the earth.

These two external efforts, central forces (apparently) between earth &sun and between earth & moon, are independent of each other. Central forces, due the presence of both the sun and the moon, act evenly and continuously on the earth to provide external linear efforts on the spinning body of earth. Each of these central forces, independently, transforms the shape of earth to increase its diameter along the directions of their actions, each effort in its own direction of action. Since the centre of earth's orbital path is far out of its body, curvature of its orbital path is small. For calculations, small part of earth's orbital path can be assumed as a straight line. Hence, tides due to an external effort, on both sides of earth, are approximately of the same height from a common reference. Small differences in their heights due to curvature of orbital path are ignored here. Variations in the parameters of tides, due to eccentricity or inclination of orbital paths, are also not considered.

For convenience, we regard shape of earth as a spheroid or a sphere. All cross sections of earth, perpendicular to its spin axis, are considered as perfect circles. In order to account for the differences in diameters of earth's cross sections due to tides, length of earth's diameter is assumed to be of a mean length and earth's datum shape is set as a sphere. Water levels on earth's surface are then related to this datum.

Equations in above given paragraphs are true for a spherical macro body, covered evenly with deep fluid matter. However, the earth in its nature has an uneven surface of landmasses and oceans.

Tidal effects felt by rigid land mass and fluid oceans are slightly different. Ocean water conforms to tidal effects freely whereas the landmass does it reluctantly. This tends to create level differences of surface points with respect to a reference. Gravitational actions due to earth's rest mass try to overcome the level differences of its surface points and create superficial flow of water bodies from one place to another, locally. However, there is no overall displacement of water bodies along with the tides. If water body was to move to create tides, there would have been a constant westward flow of ocean water (at least, in cases of bodies with no land masses to break the flow). Tendency of such flow is not observed on earth.

When the earth as a whole is considered, it may appear that the crests and troughs of the large-scale traveling wave system comprising the tides strive to sweep continuously around the earth, following the relative position of the moon (and the sun). This is mere appearance due to the motion of the observer in the opposite direction. While the earth spins, its shape remains steady in space. An observer, static with respect to earth's surface, moves through high and low tide regions in easterly direction and experience the feeling of tides traversing him in opposite direction. Changes, the observer experiences, are not caused by lateral displacement of the water body but due to vertical changes in the shape of the planet. As there is no relative flow of ocean water from one part of earth to another or around the earth's core body, laws of fluid dynamics do not apply to tides. Since there is no relative linear motion between the water body and the land mass at the ocean floor, there is no frictional effect at the ocean floor. The assumption that earth's spin speed slows down due to such friction – tidal drag – is baseless. In fact, earth's orbital motion has an accelerating effect on its spin motion.

Not all natural phenomena that cause temporary rise or fall of water level in the ocean can be interpreted as tidal effects. Tides are the rise and fall of a point on the surface of a spinning macro body, with respect to a datum, in the direction of or away from an external (linear or spin) force acting on it. Changes in the surface level, due to relative displacements of matter content of the body can not be attributed to tides.

Since the tides are formed due to acceleration of the spinning body rather than due to change of its state of motion (change in speed), no work is expended for the creation of tides. Only the shape of the body is changed. Mere (temporary) changes in shape or direction of motion do not constitute work. To change the shape of the body, original work existing within the body's matter field is re-deployed during the process of investment of work (acceleration) by an external effort. No energy is used from any source to produce tidal effects on a spinning macro body. Hence, tidal effects cannot do any work. Energy from other sources (like gravity, during changes in levels of water bodies) may be derived to do other works during tides. Work invested by external effort in a planetary body's matter field is used solely to change the state of the body's motions by modifying its linear or spin speeds in its orbital path. Central force on a planet produces the orbital motion and spin motion of the planet. Transitional-stage (acceleration period) of the central force's action causes the tide.

In case of terrestrial tides, effects due to eccentricity of orbital paths, inclination of orbital planes, topography on the ocean floor, flow of water into confining channels or nearly closed oceanic basins, dynamic considerations during local flow of water due to level difference, atmospheric conditions and contiguous current in the oceans are also needed to be taken into account. All of the above (and other less important influences) can combine to create a considerable variety (of many times) in the observed magnitudes and phase sequence of the tides - as well as variations in the times of their arrival at any location. Of a more local and sporadic nature, important meteorological contributions to the tides are know as 'storm surges'. They are caused by a continuous strong flow of winds either onshore or offshore. They may superimpose their effects upon normal tidal actions to cause variations in the magnitudes of high or low tides. High-pressure atmospheric systems may depress the tides and deep low-pressure systems may cause them to increase in height. Higher inclination of lunar orbit makes large variety between tides at the equatorial region and higher latitudes.

Calculating the magnitude of terrestrial tides by using equation (7/42) and putting the value of apparent gravitational attraction from equation (7/10), we can see that the magnitude of lunar tides are about 2.3 times greater in magnitude than solar tides on earth. This is irrespective of the fact that by conventional equations, earth's apparent gravitational attraction towards the sun is about 178 times greater than its apparent gravitational attraction towards the moon.

7.6.2. Direction of tides:

It is often observed that zenith points of terrestrial tides do not coincide with the local meridian where the sun or moon is present. This is usually attributed to friction between water and landmasses of earth. It is not so. Even if the planet is wholly fluid, similar changes in the direction of tides will appear. Change in the direction of tides is caused by the fact that direction of action of an effort need not always be wholly in the direction of its application, as explained in the parent book. Changes in the direction of tides are local phenomenon related only to the parameters of the spinning body. Hence, magnitudes or sources of external effort or the parameters of the source-bodies do not affect the changes in the apparent deflections of tides from local meridian.

Sun, the central body of earth's orbital motion, moves at an absolute linear speed in space. Depending upon the position of earth on its orbital path, with respect to the sun, earth's absolute linear speed varies. This makes corresponding changes in the magnitude of angular shift of the zenith point of the terrestrial tide from the local meridian. Angular shift of the tidal zenith point depends only on the parameters of earth. Hence, irrespective of the source of external effort, acting on the earth, all terrestrial tides are shifted identically. Magnitudes of deflection depend on earth's spin speed and its location on its orbital path with respect to the source body of external effort.

Distortions on matter field latticework squares in a rotating spherical body, angularly deflects mean normal of the latticework squares from radial lines of the body. Angular displacement of the mean normal of latticework square to the tangent at the periphery of a planetary body's matter field causes the tides to lead or lag from local meridian.

Magnitude of angular shift:

Figure 7U shows one quarter part of a 2D energy field square situated at the zenith point 'A' (on the line of application of the external linear effort on a spinning body, refer figure 7T) on the equatorial perimeter of a spherical body. AD is one of the quanta of matter forming the matter field square. Work (distortion), causing body's rotary motion, displaces junction point 'D' towards 'O'. End 'D' of the quantum of matter is displaced to 'C' through a distance proportional to tangential speed, 'v', of a matter particle in that region. As and when the spinning body develops additional linear speed equal to 'V', point 'C' is displaced by additional distance, CB, proportional to 'V'. Speed, 'V', considered here is of absolute nature. It is with respect to the 2D energy fields in space.

In the figure 7U;

$$\angle OAD = \pi/4 \text{ rad, } OA = R, BC = V, CD = v,$$

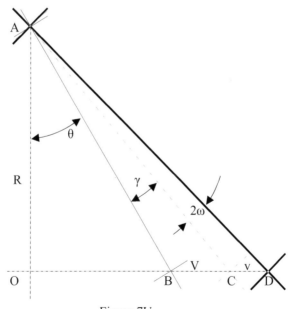

Figure 7U

Where v is the tangential speed of the surface particle with respect to the (axis of the) spinning body and V is spinning body's absolute linear speed.

Let the spin speed of the body is equal to ω rad/sec. Spin speed of the body is equal to the angular deflection of median between side AD and opposite side of the matter field square, quarter part of which is shown in figure 7U. Matter field-distortion, corresponding to the spin motion of the body, deflects only the arm AD of the matter field latticework square. The arm of the latticework square, opposite to AD, is not deflected. Therefore, in order to produce a deflection of angle ω of the median line between the arm AD and its opposite arm, side AD has to have a deflection of 2ω.

$$\angle CAD = 2\omega, \quad \angle OAC = \pi/4 - 2\omega,$$

$$\text{Tan} \angle OAC = \text{Tan}\left(\pi/4 - 2\omega\right) = \frac{R - v}{R}$$

$$R = \frac{v}{1 - \text{Tan}\left(\pi/4 - 2\omega\right)}$$

$$\text{Tan }\theta = \frac{R - V - v}{R} = \frac{\dfrac{v}{1 - \text{Tan}\left(\pi/4 - 2\omega\right)} - V - v}{\dfrac{v}{1 - \text{Tan}\left(\pi/4 - 2\omega\right)}}$$

$$= 1 - \frac{(V + v)\left(1 - \text{Tan}\left(\pi/4 - 2\omega\right)\right)}{v}$$

$$\theta = \text{Tan}^{-1}\left\{1 - \frac{(V + v)\left(1 - \text{Tan}\left(\pi/4 - 2\omega\right)\right)}{v}\right\} \text{ rad/sec.} \quad \textbf{(7/43)}$$

Total deflection of the arm due to linear speed and spin motion,

$$\gamma + 2\omega = \angle OAD - \angle OAB$$

$$= \frac{\pi}{4} - \mathrm{Tan}^{-1}\left\{1 - \frac{(V+v)\left(1 - \mathrm{Tan}\left(\pi/4 - 2\omega\right)\right)}{V}\right\}$$

Since there is no deflecting motion on opposite arm, it is not deflected. Hence, deflection of the median line between AD and the opposite arm is half of deflection of arm AD.

Deflection of symmetry of matter field square;

$$D_s = \frac{\angle BAD}{2} = \frac{\gamma + 2\omega}{2}$$

$$= \frac{\dfrac{\pi}{4} - \mathrm{Tan}^{-1}\left\{1 - \dfrac{(V+v)\left(1 - \mathrm{Tan}\left(\pi/4 - 2\omega\right)\right)}{V}\right\}}{2} \quad \text{rad/sec.} \quad \textbf{(7/44)}$$

An external linear effort, acting perpendicular to the geometrical tangent at a point on the surface of a spherical spinning macro body, acts at an angular deflection from the true perpendicular to the tangent to the matter field of the body. Magnitude of the angular deflection is given by the equation (7/44). Therefore, as far as the matter field of the spinning body is concerned, its zenith points, where the external efforts act, are away from the geometrical zenith points, determined by the observer, by angular deflection of magnitude as given by the above equation (7/44).

Zenith points of terrestrial tides are shifted from local meridian at any point on earth's equator (And similarly for all surface points in other planes parallel to equatorial plane). Since the shift is a function of earth's parameters, magnitude of shift on earth is identical for both lunar and solar tides. They are apparently related to the sun or moon only because the central forces, causing the tides, are developed in relation to these bodies. Magnitude of shift is with respect to earth's centre of rotation. An observer on earth views the tides with respect to earth. The observer is also moving with the earth at its absolute linear speed. Hence, displacement of tide from the local meridian with respect to earth's spin axis is also of the same magnitude.

Magnitude of deflection of tide from the local meridian is related to the absolute linear speed of the planet and the relative direction of the central force to the direction of planet's absolute linear motion. All bodies in the solar system move with average linear speeds equal to sun's linear speed in its orbital path about the centre of galaxy. Disregarding its spin motion, a stable galaxy can be considered as static in space, with respect to neighboring galaxies. If the deflection of tides from local meridian in a place (of uniform ocean depth and far away from landmasses) can be accurately measured, absolute speed of earth in space can be determined.

Direction of angular shift:

In figure 7V, rectangles in dashed lines, A and C, represent distorted matter field squares at the local meridian in a linearly moving spinning macro body. Direction of spin of the body, as shown in the figure, is anti-clockwise. Since the body has only one combined motion (linear motion along a curved path about the centre of rotation of the body) shown by the thick curved arrow and the curved path's centre of curvature is very far from the macro body, all matter field squares in the body are, more or less, distorted identically. Shorter arrows represent two sets of external linear forces applied on the body in two directions, upward and downward. Other two matter field squares, B and D, are displaced from the local meridians. D is displaced in the same direction of body's spin motion, from C and B is displaced in opposite direction to body's spin motion, from A. Central force (a push effort), acting on the spinning macro body, is towards the central body. Matter field-distortions of a spinning macro body are not be symmetrical about the direction of an external effort, acting evenly on it. Highest magnitude of tides occur at the point, where distorted matter field squares are symmetrical about the direction of external effort.

Therefore, an observer standing on a planet sees the tide either leads or lags the local meridian. [Local meridian is the location on the surface of planetary body, which is directly under the celestial body, considered]. This is mere appearance. Zenith points of tides always take place, where (distorted) matter field squares are symmetrical about the line of external efforts. When the observer is directly under the central body, distorted matter field squares at the local meridian of the spinning planetary body are not symmetrical with the line of action of external effort (central force). Matter field squares become symmetrical about the line of effort only when the observer is away from the meridian, facing the central body. As far as the matter field of the planet is concerned

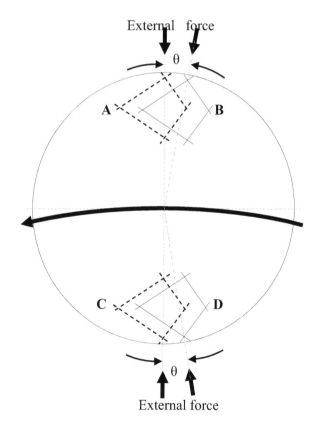

Figure 7V

this is the true vertical direction to planet's surface. The observer sees this as an angular shift in the place of occurrence of tide from the local meridian occupied by him.

Let us consider action of downward external effort on the matter field square 'A' in figure 7V. At its position at 'A', action of external effort is asymmetrical on its arms. Matter particles in contact with this matter field square experience anti-clockwise deflection in addition to downward linear motion. Maximum tidal effect occurs where no part of the external effort is used for angular deflection of particles' paths. Since the external effort is not fully effective at 'A', in downward direction, high tide cannot occur at this position (local meridian). As the planetary body rotates, observer at A is carried forward. After some time, matter field square situated at 'B' comes to occupy position in line with the external effort. This is shown in the figure by a shift in the direction of external effort, instead of displacement of the matter field square. Here, the arms of matter field square 'B' are symmetrical about the direction of external effort. Whole of the external effort acts along its direction of application. No part of the external effort is used to deflect matter particles, angularly. Hence, high tide occurs at this point. By the time matter field-square 'B' comes in line with the external effort, local meridian of the observer has moved ahead in the direction of spin of the body. Therefore, when the external effort is applied towards the centre of resultant motion of the spinning body, observer notices that the high tide lags behind the local meridian occupied by him.

Consider action of external effort in the opposite direction, away from the centre of resultant motion of the body. At local meridian occupied by the observer, as shown by matter field square 'C', its arms are asymmetrical to the external effort. Hence, the high tide cannot occur at this point. Matter field square at position 'D' has its arms symmetrical about the external effort. Tide is highest when the meridian at position 'D' is facing the central body. Local meridian occupied by the observer is behind the meridian where high tide takes place. When the external effort is applied in the direction away from the centre of resultant motion of the spinning body, to the observer the tides appear to lead the local meridian (where the central body appears to him).

Figure 7W shows the orbital path, GA'CBDAF, of earth about the sun. Thick arrow T_1R_1 shows the resultant direction of earth's motion (linear motion + spin motion) along its median path. Block arrows show the direction of central force between earth and the sun. Central force on a planet is a push effort, applied from the side away from the central body towards the central body. Small circles show earth's spin direction and arrows across the circles shows the direction of solar tides with respect to local meridian in each quarter of the orbital path. Arrows in dotted line show the direction of earth's absolute linear motion with respect to sun's path. In real sense, deflection of solar tides with respect to meridian is always as shown in this figure and as explained above. In positions P and R, solar tides are deflected westward from local meridian and in positions S and T, solar tides are deflected eastward from local meridian.

During the half-yearly period, when the earth is nearer to outer datum point in its orbit (it is outside its median path - with respect to sun) the earth is farther than the sun from the centre of sun's path. The central force is applied from the side away from the sun towards the centre of earth's combined motion. [Central force is a push-effort applied on earth's farther side, towards the sun]. During this six-month period, solar tides tend to lag behind the local meridian of the observer. Tides will appear to the west of local meridian (they appear earlier than the sun itself reaches the local meridian).

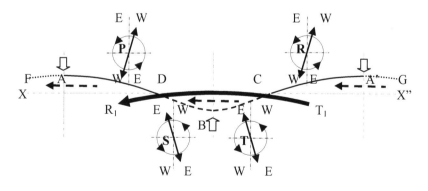

Figure 7W

During the half-yearly period when the earth is nearer to inner datum point in its orbit (it is inside its median path - with respect to sun) the earth is nearer to the centre of sun's path. During this time, the central force is applied away from the centre of earth's resultant motion. [Central force is a push-effort applied on earth's farther side towards the sun]. During this six-month period, solar tides tend to lead the local meridian of the observer. Tides will appear to the east of local meridian (they appear later than the sun itself has crossed the local meridian).

Directions of lunar tides are also similar. Figure 7X shows earth's orbital path, T_1R_1 and the moon moves along the orbital path GA'CBDAF about the earth. Block arrows show the direction of central force between moon and earth. Central force on a planet is a push effort, applied from the side away from the central body (in this case the moon) towards the central body. Small circles show earth's spin direction and arrows across the circles shows the direction of lunar tides with respect to local meridian in each quarter of moon's orbital path. Arrows in dotted line show the direction of earth's absolute linear motion with respect to moon's path. In real sense, deflection of lunar tides with respect to meridian is always as shown in this figure and as explained above. In positions P and R, lunar tides are deflected eastward from local meridian and in positions S and T, lunar tides are deflected westward from local meridian.

During the half-monthly period, positions S & T, when the

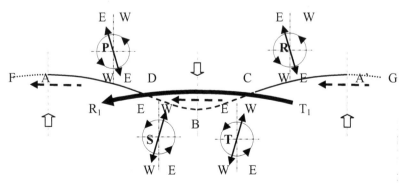

Figure 7X

moon is near the inner datum point in its orbit about the earth (it is within earth's orbital path - with respect to earth) the moon is nearer to the centre of earth's resultant motion. The central force on earth is applied towards the centre of earth's resultant motion. [Central force is a push-effort applied on earth's side away from the moon towards the moon]. During this half-monthly period, to the observer, the lunar tides tend to lag behind the local meridian. Tides will appear to the west of local meridian of the observer (they appear earlier than the moon itself reaches the local meridian).

During the half-monthly period, positions P & R, when the moon is near the outer datum point in its orbit about the earth (it is outside earth's orbital path - with respect to earth) the moon is farther from the centre of earth's combined motion. The central force is applied in a direction away from the centre of earth's combined motion. [Central force is a push-effort applied on earth's side away from the moon towards the moon]. During this half-monthly period, lunar tides tend to lead the local meridian. Tides will appear to the east of local meridian of the observer (they appear later than the moon itself has crossed the local meridian).

If there are more than one central force, acting on a spinning planetary body, each central force produces its own set of tides, independently. If the directions of these tides are near, they will create resultant tides, which are arithmetical sum of independent tides on the body. This summation gives rise to spring and neap tides on earth. This effect is the greatest when the Moon and Sun are in a straight line with the Earth, called 'syzygy', which occurs during a Full Moon and New Moon (during Lunar and Solar Eclipses). To an observer of earth, earth's orbital path does not appear as a wavy line but it is observed as circular/elliptical path around the central body. This appearance (different from real condition) further change the directions, how the tides appear to an observer.

Direction of Solar tides:

An observer on earth judges the orbital motion of the earth about the sun as he sees it. In the past, to the observer, the sun appeared to move around the earth in westerly direction. This notion

Chapter 7. **GENERAL**

was later changed to the motion of the earth, around the sun in easterly direction. Although, no free body can orbit around another moving body, the notion of earth orbiting around the sun in easterly direction is still maintained. Adherence to this incorrect belief is the cause of many misunderstandings in celestial mechanism. In order to satisfy this belief, directions of certain motions are arbitrarily changed by the learned observer.

In the figure 7Y, XX" shows part of sun's path. Earth's orbital path is shown by wavy line GA'CBDAF. A and A' are outer datum points (points in the orbital path, where absolute linear speed of earth is highest) and B is the inner datum point (points in the orbital path, where absolute linear speed of earth is lowest). For the time being, it is assumed that the central force, acting on earth, is always (nearly) perpendicular to its orbital path. Arrows in dotted lines show relative directions of linear motion of earth, in each quarter, with respect to sun. As the earth moves in its orbital path, from inner datum point, B to outer datum point, A, it is behind the sun. Central force between earth and sun accelerates the earth in forward direction towards X. Relative direction of earth's linear motion is in the same direction as that of the sun's linear motion. Directions of earth's spin and the deflection of tide from local meridian, as shown by P and S, are same as derived in above section. During earth's travel from D to A, solar tides are deflected westward from local meridian, as shown by position P. And during earth's travel from B to D, solar tides are deflected eastward from local meridian,

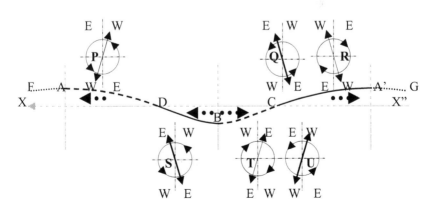

Figure 7Y

Chapter 7. GENERAL

as shown by position S in figure 7Y.

North-south directions in space are oriented with respect to earth's spin axis and this orientation is considered true throughout the space. Unlike north-south directions, east-west directions have no definite orientation in space. These are indicated by the direction of motion of a surface particle on earth. Since the spherical earth has a spin motion, relative to space, east-west directions depend on the motion of a location of surface point on earth.

From A' to B, in figure 7Y, earth is in front of sun and hence it is retarded in its linear motion. Sun appears to advance towards X while earth appears to move (relatively) in the opposite direction, shown by arrows in dotted line. This assumption is able to change the shape of earth's orbital path from a wavy line to a path around the sun. By this consideration, earth appears to move from B to A'. Direction of earth's real motions is reversed by this supposition, as shown by positions T and R. By doing so, not only earth's linear motion but its spin motion is also reversed. Change in the spin direction is against what is observed. Here, again, we resort to one more change (Or undo part of the change of direction attempted earlier). East-west directions are changed back to suit the observation related to earth's spin motion. This is shown in positions Q and U. By doing so; in position Q, relative direction of solar tides have changed easterly and in position U, relative direction of solar tides have changed westerly with respect to local meridians. Although the reality is different, this is what we observe and believe to be true.

Summarizing the above points;

- From A' to C, solar high tides occur before the sun itself reaches the local meridian of the observer. Solar tides are deflected in easterly direction as shown in position Q.

- From C to B, solar high tides occur after the sun itself crossed the local meridian of the observer. Solar tides are deflected in westerly direction as shown in position U.

- From B to D, solar high tides occur before the sun itself reaches the local meridian of the place. Solar tides are deflected in easterly direction as shown in position S.

- From D to A, solar high tides occur after the sun itself crossed the local meridian of the observer. Solar tides are deflected in westerly direction as shown in position P.

High tides on the opposite side of earth also appear correspondingly.

Direction of lunar tides:

Figure 7Z shows the relative orbital motion of moon with respect to earth when earth is at and near the outer datum point in its orbital path. Moon travels in the wavy path, GA'CBDAF, while the earth moves along X"X in its orbital path about the sun. To an observer on earth, the moon appears to move around the earth. Earth's motion, relative to the moon, appears in directions as indicated by dotted arrows. From A' (full moon) to B (new moon), the moon is in front of earth and the earth accelerates due to the central force in the direction of moon's motion. Earth appears to move towards the moon, which is in front of earth. Both bodies appear to move in the same direction. Directions of deflections of the tides are as explained above, with respect to figure 7X and at positions Q and U.

Between B (new moon) and A (full moon), the moon is behind the earth and the earth is decelerated due to the central force in opposite direction to moon's motion. Earth appears to move towards the moon (in opposite direction to moon's motion), which

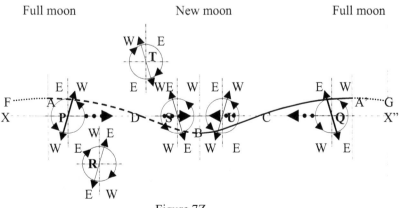

Figure 7Z

is behind the earth. They appear to move in opposite directions. This appearance creates the apparent orbital motion of moon around earth. Directions of earth's real motions are reversed to suit observation as shown by circles, at positions T and R, in dotted lines. Further, direction of earth's spin motion is reversed to suit the observation as shown in circles at positions P and S. Resulting appearances of deflections of lunar tides are shown by arrows across circles at positions P, S, U and Q.

Summarizing the above;

- Between new moon at position B and first-quarter phase at position D, lunar high tides occur before the moon itself reaches the local meridian of the place. Lunar tides are deflected in easterly direction as shown in position S.

- Between first-quarter phase at position D and full moon at position A, lunar high tides occur after the moon has reached the local meridian of the place. Lunar tides are deflected in westerly direction as shown in position P.

- Between full moon at position A' and third-quarter phase at position C, lunar high tides occur before the moon itself reaches the local meridian of the place. Lunar tides are deflected in easterly direction as shown in position Q.

- Between third-quarter at position C phase and new moon at position B, lunar high tides occur after the moon has reached the local meridian of the place. Lunar tides are deflected in westerly direction as shown in position U.

High tides on the opposite side of earth also appear correspondingly.

Effect of orbital motion on direction of tides:

Magnitudes and directions of shift of tides, as explained above, are satisfied only when the central force is perpendicular to earth's orbital path. This can be so only under the condition that earth's orbital path is circular around the sun and moon's orbital path is circular around the earth. From the above figures, it can be seen that earth's orbital path zigzags about sun's path (and moon's orbital path zigzags about earth's orbital path) and the direction of central forces with respect to orbital paths changes through a full circle during every fractional orbital period. Central force is perpendicular to the orbital path only at two points (at outer and inner datum points) in an orbit. At all other points in the orbit, angles between orbital path and the central force vary between 0° and 90°.

Direction of earth's orbital path deflects to a maximum of about 6° from its median path on either side. Median path happens to be the sun's path. Earth's matter field squares deflect through a maximum of 6° on either side about median deflection. Deflection in earth's matter field squares due to its orbital motion enhances the deflection of high tides from local meridian, occupied by an observer. Accordingly, depending upon the location of earth in its orbit, deflection of solar high tides from local meridian of an observer increases up to about 9°, where angular deflection of orbital path from median path is highest. (At the point, where earth crosses sun's path - in case of solar tides and at the point, where earth crosses moon's path - in case of lunar tides, in space). At these points, only one-third of deflection is caused by the asymmetry in matter field squares of the earth and the rest is caused by curvature of earth's orbital path.

Magnitudes and directions of shift of lunar tides also vary between a minimum to a maximum, during one solar year. Magnitude of angular shift of lunar tides varies from one lunar month to another lunar month, completing one cycle in one solar year. Magnitude of deflection of terrestrial tides varies within an angular sector of on either side of local meridian. Magnitude of deflection, at any time, depends also on many other factors like; the locations of the earth and the moon in their respective orbital paths, relative direction of their orbital motion, etc.

Tidal effects on a spinning body take place separately in each plane perpendicular to its spin axis. Acceleration (linear or rotational) causes tides on a spinning body. There are neither displacements of body parts or flow of ocean water during tidal formation. Superficial flow of ocean water during tides is caused by effective level differences of earth's surface. Tidal drag on earth's solid body is a fallacy. Absolute linear motion of a spinning planet produces angular shift of tides from local meridian facing its central bodies. Directions of apparent shift of tides do not conform to their real deflections due to direction of central forces. Curvature of earth's orbital path has greater effect on displacement of tide from local meridian than earth's absolute linear speed. Phenomenon of tides on planets should be interpreted on facts rather than on their appearances.

7.7. Galactic repulsion:

Whether it is due to mutual attraction or due to changes in fields about a body (changes in space-time continuum), it is a fact of observation that all matter bodies in nature have a tendency to approach each other. Discovery of this fact has raised further logical questions. It is observed that matter bodies are present everywhere in space. If they have a tendency to approach each other, at some stage of universe, all matter bodies in nature should form a single matter body and presence of matter will be concentrated at or near a point. This is not logical and contrary to present state of universe as observed. However, this illogical thought has led to many irrational theories.

Thus, it has become necessary to discover a logical mechanism that keeps total matter content of the universe widely spread throughout the universe. In the past, many theories were devised to justify widely-spread presence of matter bodies, irrespective of apparent attraction between them. Unfortunately, none of them gained wide acceptance as a logical theory.

In this concept, certain distortions in the universal medium cause apparent gravitational attraction between matter bodies. If these distortions can be modified by another natural phenomena, apparent attraction between matter bodies can be counteracted to keep matter bodies of certain sizes, away from each other. At the same time, such neutralization of apparent gravitational attraction should not be effective between bodies of differing and small sizes. It is the nature and magnitude distortions in the universal medium (2D energy fields), which dictates the nature of apparent interactions between matter bodies.

Speed of light is the ultimate linear speed of matter bodies in nature. As linear speed of a macro body approaches linear speed of light, it breaks down to its constituent fundamental particles and primary particles. At the linear speed of light, only matter particles that can survive are the photons. Beyond this speed, no matter body can be moved because that is the ultimate speed the 2D energy fields can provide. Any attempt to increase linear speed of a photon increases its matter content (frequency) rather than its linear speed.

Biton is a primary particle made up of a binary union of two identical photons. Constituent photons of a biton maintain their linear motion at critical linear speed in a common curved path about a common centre. Simultaneously, matter cores of constituent photons of a biton spin in phase about a common axis passing through biton's centre. Bitons are unable to move at any appreciable linear speed in planes of their existence. At higher linear speeds, 2D energy fields reorient all bitons to move in planes perpendicular to their planes of revolution.

Due to angular nature of associated distortion field of a primary particle (biton), each biton has a primary electric field. An electric field is an angular distortion field in the 2D energy fields. Face of an electric field, where representative lines of force appear in clockwise direction is positive electric charge. Face of an electric field, where representative lines of force appear in anti-clockwise direction is negative electric charge. Every electric field has both positive and negative electric charges.

Combination of electric fields in proper array creates magnetic fields. Magnetic fields are linear distortion fields in 2D energy fields, where lines of force of distortions are of linear nature. An electric field with small curvature of lines of force acts as a magnetic field.

A free electric field-producing element, moving in a magnetic field with a gradient (gradually varying in strength), tends to reorient itself such that it is in attractive (interactive) phase towards higher-density region of the magnetic field.

A very large inter-galactic cloud, during its condensation period, may be fragmented into many smaller clouds by uneven distribution of its matter content and by spinning motion of the cloud, as is envisaged in 'Nebular hypothesis'. These smaller clouds further condense into separate bodies but simultaneously being constituents of the same group. In this case, the total matter content of the combined body is distributed over a wider region and hence there is no concentration of its matter content in a place. Photons, escaping from the region of the cloud are not slowed down very much and hence these types of groups of bodies, called 'galaxies', are visible to outside observers within the universe.

A very large (single) macro body of matter content, comparable to a galaxy, is a 'black hole'. A black hole could also be a member-body of a galaxy. Main difference between black hole and a galaxy is in the distribution of their matter content. In a black hole, whole of its matter content is concentrated in a single body but the total matter content of a galaxy is distributed over a wide region in space in the form of small bodies and dust clouds. Unusual concentration of matter in any part of a galaxy may cause that part of galaxy to form into a black hole.

Galaxies spin as a single body, over and above the spinning and revolving motions of local bodies within the group. Each part of the galaxy and the galaxy as a whole body develop their spin motion independently. A galaxy develops it spin motion due to its gravitational collapse and uneven distribution of its matter content. Spin motions of other bodies are developed by their gravitational collapse as well as by apparent interactions between them. A galaxy may contain one or few black holes, millions of stars, planet-sized bodies and smaller bodies. It also has assorted very small sized bodies along with dust clouds.

Due to the very large size of a galaxy and its high spin speed, linear speed of any matter particle towards the edge of the galaxy is extremely high and is comparable to the speed of light (photons). No 3D matter particles larger than the photons and the bitons (with their planes perpendicular to the direction of motion) can survive at this speed. Therefore, all matter bodies in the region along the periphery of a galaxy find themselves disintegrated in to their constituent bitons. These bitons orient themselves to minimize resistance to their motion, from the 2D energy fields, by having their planes perpendicular to their line of motion. This is purely a mechanical action so that the moving bitons experience minimum resistance from the 2D energy fields. Outer edge of a galaxy is filled with independent bitons, moving in this state. Peripheral region of a galaxy, occupied by free bitons, is the 'halo' of galaxy. Apparent attraction due to gravitation between bitons in halo and matter content of the galaxy is balanced by the apparent centrifugal action on the bitons due to their linear motion in curved paths around the galaxy.

Each biton has a primary electric field. Orientation of a biton along the periphery of the galaxy is a mechanical activity. Therefore, these bitons are randomly oriented in the beginning. They could be oriented in either of the two ways. Their electric charges could be in phase or out of phase with a reference. Primary electric fields of equal numbers of bitons, which are out of phase with each other, neutralize their primary electric fields. There will be some surviving primary electric fields, in the halo, which produce a resultant electric field in any one direction. Primary electric field acts within zilch force distance with other distortion fields, in their immediate neighborhood. Free primary electric fields in the halo, whose primary electric fields are not neutralized, together, make a resultant electric field in the shape of a torroid along the outer edge of the galaxy. Resultant electric field being large (of low curvature), acts outside its zilch force distance and hence behave like a magnetic field. Thus, there is a strong magnetic field around the edge of a (spinning) galaxy, perpendicular to its plane of spin. Lines of forces of magnetic fields at two places on the periphery of a galaxy are shown in figure 7AA. Directions of

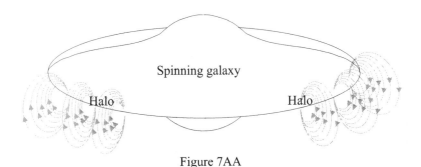

Figure 7AA

these magnetic fields, appearing in the halo on the periphery of the galactic disc, are with respect to an external reference. Each biton, contributing to this magnetic field, is also capable to interact on its own with any other external distortion fields.

Galaxies in space are also under gravitational influence. They apparently attract each other due to gravitation. Galaxies tend to move towards each other under apparent gravitational attraction. If they are near enough, magnetic fields about their halos interact with each other. There are two possibilities. Their magnetic fields can be in repulsive phase or in attractive phase with each other. Both of the galaxies, moving towards each other under the apparent attraction due to gravitation, have strong magnetic fields about their periphery, perpendicular to their plane of spin.

As their magnetic fields starts to interact, each of the primary electric fields of free bitons, present about the rim of galactic periphery, is also interacting with the magnetic field of the other galaxy in its own capacity. They are able to act on their own because each biton is an independent body and is not bound to any other particle, mechanically or otherwise. Because of the great distance between the galaxies, there is a gradient in their magnetic field strength along the line joining their centers. Magnetic field gradient of one galaxy affects the primary electric fields of bitons of the other galaxy. Each of the primary electric field tends to re-orient itself so that it is in attractive phase towards the higher-density region of the magnetic field of the other galaxy. Many of the bitons succeed to re-orient themselves in this way.

Ability of a biton to re-orient itself depends on the relative strength of re-orienting effort with respect to the aligning effort due to its linear speed. Re-orientation of a biton does not always mean that the biton is turned through 180 degrees, but it is turned by angle enough to make a change in the sense of its primary electric field with respect to external distortion fields. Figure 7AA shows magnetic fields produced at two different places in the halo of a galaxy, facing two external galaxies. They are of different polarities.

Two parallel and unidirectional distortion (like-magnetic) fields repel each other. Let the magnetic lines of forces of two approaching galaxies are of the same polarity. Their lines of forces are parallel and unidirectional. Hence, the galaxies apparently repel each other. If their repulsion is strong enough, the galaxies come to stay away from each other, at a place where the apparent attraction due to gravitation and the apparent repulsion due to their magnetic fields balance each other. Since the galaxies are spinning bodies, this balancing is a dynamic action. As the galaxies turn, nearest points on their peripheries facing each other, change. Magnetic field strength at these points also may be different. Therefore, variation in the strength of galaxies' magnetic fields has to be continuously updated to maintain the required balance.

If their apparent repulsion is not strong enough, due to low magnitude of their magnetic fields, approaching galaxies may continue to move towards each other. Now, the free bitons, in their halos, moving along the periphery of each galaxy is carried into the denser part of magnetic field of the other galaxy. Magnetic fields of the galaxies have higher density in a direction towards galactic center. Therefore, these bitons (disregarding their movements along the periphery of the galaxy) are moving towards a region of higher-density magnetic field. Bitons moving towards higher-density magnetic region tend to re-orient themselves, such that their primary electric field is in attractive phase with the higher-density magnetic region. Bitons in both the galaxies tend to re-orient and many of them will succeed. If the approaching speed of the galaxies are faster, more bitons are re-oriented and at a higher rate. Re-orientations of these bitons strengthen magnetic fields of both the galaxies to increase their magnetic field strength and their mutual apparent repulsion.

Two parallel distortion fields (magnetic fields) in opposite directions attract each other. Let the magnetic lines of forces of the two approaching galaxies are of opposite polarity. In this case, lines of forces are parallel but in opposite directions. Hence, the galaxies apparently attract each other due to their magnetic fields. This apparent attraction is assisting the apparent attraction already existing between them due to gravitation. The galaxies are bound to move in a collision course at an accelerating pace.

Since the magnetic fields of approaching galaxies (let them

be galaxy 'A' and galaxy 'B') are in opposite directions, they tend to neutralize each other. Only the resultant of the two magnetic fields is left over. Therefore, for the time being, we will consider that galaxy 'A' has greater magnetic field compared to galaxy 'B'. Resultant magnetic field of their combination belongs to the galaxy 'A', whose magnetic field is of greater strength. As the galaxies move towards each other, free bitons on the rim of galaxy 'B' are carried into the magnetic field of galaxy 'A', in a direction towards high-density region of the magnetic field. These bitons tend to re-orient so that they are in attractive phase with the region of higher-density magnetic region. Many bitons succeed to re-orient themselves. This reduces magnetic field of galaxy 'B', which was in opposite direction to the magnetic field of galaxy 'A'. Process of re-orientation of the bitons will continue and gradually the galaxy 'B' will develop a magnetic field, which is in the same direction as that of galaxy 'A'. Now, magnetic fields of both galaxies are in repulsive phase. Strength of the apparent repulsion between the galaxies will be adjusted in due course of time as described earlier.

By the re-orientation of the bitons in their halos, in this way, the resultant electric field/magnetic field of parts of both galactic peripheries, facing each other, have now become in repulsive phase with each other. Thus, the galaxies are prevented from coming in to colliding distance, irrespective of their relative direction of spin. Factors controlling this phenomenon are the direction of the magnetic field of one galaxy and the direction of orientation of the free bitons in the other galaxy. Because of this action, it is possible for a galaxy to have different directions of its magnetic fields at different places in its halo around its periphery, facing other galaxies.

Only factor, producing this kind of apparent repulsion between the galaxies, is the ability of free bitons in the halos to re-orient themselves, irrespective of their direction of motion. Hence, any two large bodies with similar high-speed spin and with free bitons at their periphery can develop magnetic fields to produce apparent repulsion between them. Relative directions galactic planes or their shapes do not affect this phenomenon. Any two galaxies (even if their direction of approach is along their spin axes) are prevented from approaching each other within collision distances. They may collide only in accidental situations, which are most improbable.

If sufficient time is not available to create enough apparent repulsion between the galaxies, they will collide into each other. Usually, magnetic interactions between spinning galaxies keep them at a definite distance from each other. Distance between two galaxies, in stable state, depends only on their matter contents. That is, distance between two galaxies is proportional to the apparent gravitational attraction between them. Their magnetic field strengths are automatically corrected to keep this distance.

Macro bodies, smaller than a galaxy, do not have this protection. Here, it is the size of the macro body, what counts and not its matter content. Many of the smaller macro bodies are spinning and have magnetic fields of their own but they do not apparently interact in this way for two reasons. First, their peripheral speed is too slow to have free bitons to form halos along their periphery. Secondly, magnetic field-producing elements (free bitons or other fundamental matter particles at their periphery) are not free, as to reorient themselves under the influence of an external magnetic field. As a result, these macro bodies approach each other under apparent gravitational attraction between them to collide into each other or to be captured-in, to form union of multi-body system.

This phenomenon gives galaxies their ability to exist independently and static in space (relative to other galaxies). Hence, wherever in space we look, we may find galaxies there. Stable galaxies in space constitute matter-world to us. This matter-world, on a large scale is in steady state and static. However, macro bodies are not perpetual. Locally in any part of a galaxy or the galaxies themselves are destroyed and rebuilt in cyclic manner.

7.8. Gravitational collapse of large bodies:

A photon is a corpuscle of light or any other similar radiation. It is a basic 3D matter particle. By their inherent properties, 2D energy fields move a photon at the highest possible linear speed, which is the linear speed of light. This linear speed is constant for any region of space and depends only on the nature of 2D energy fields in the region. An attempt to increase a photon's linear speed compels the 2D energy fields to increase photon's matter content (frequency) by adding quanta of matter to its core body, rather than increase its linear speed. Similarly, an attempt to reduce a photon's linear speed compels the 2D energy fields to reduce photon's matter content (frequency) by removing quanta of matter from its core body, rather than reduce its linear speed.

Bitons are primary matter particles, constituted by two complimentary high frequency photons as a binary system. Explanations on the formation, structure and properties of bitons may be found in the parent book. Perimeter of a biton is one wave length of its constituent photons' inertial pockets. An external pressure on a biton (while forcing them towards each other) compels its constituent photons to lose matter and energy contents from them. Loss of matter contents reduce photons' frequency and thereby increase their wave-lengths. Due to increase in the wavelengths of the constituent photons, a biton increases its radial size, as it loses quanta of matter from its constituent photons. This process is heating. During heating, a macro body (generally) expands and radiates heat rays. Reverse actions take place, when external pressure on a biton is reduced, during cooling process. In certain cases, molecular arrangements of the macro body may produce anomalies in its expansion/contraction during heating/cooling.

Quanta of matter, lost from the bitons to surroundings, during heating, form disturbances in 2D energy fields. If supply of free quanta of matter is sufficient, they form larger disturbances, which may be converted to independent photons and radiated from the region. Depending on the matter contents of photons, produced in the region, they may constitute infrared (heat), visible (light) or of higher frequency radiations. This property of bitons causes (heat) radiation from a compressed body.

Solid and liquid bodies may not be readily compressible by mechanical means. But compression of gaseous body by mechanical means easily causes the body to heat up and radiate heat rays. If the compression pressure is high enough, all macro bodies will behave the same. As compressive pressure on a macro body is increased, its constituent bitons expand and they loose matter content (with corresponding energy content). Simultaneously, matter particles of the body are brought nearer against the field forces, which are keeping them away from each other. Bringing matter particles of a macro body, increases body's internal pressure. Internal pressure of a macro body acts as external pressure on its constituent bitons.

Every matter particle in a macro body is under apparent gravitational attraction with all other matter particles in the body. This helps the body to maintain its integrity as a single body. If a macro body is large enough, apparent gravitational attraction between its constituent matter particles may bring them nearer by overcoming field forces (which keep them away from each other) to increase body's internal pressure. This phenomenon is a macro body's 'gravitational collapse'.

Even in cases of planet sized bodies, low frequency radiations from them are (correctly) believed to be the result of gravitational collapse experienced by them. However, when it comes to the cases of stars, an entirely new mechanism is assumed to be responsible for radiations from them. It is believed that fusion between atoms (of low atomic number) is responsible for radiations from stars. This need not be so. If the macro body is large enough, actions during gravitational collapse of the body will be able to raise its internal pressure to very high levels. This can cause radiations of all frequency ranges, including very high frequency radiations from them.

If a single large macro body contains sufficient matter content, as the body undergoes gravitational collapse, its internal pressure can increase to very high values. Internal pressure of a body is experienced by its constituent biton as external pressure on them. As the compression rate of the macro body increases, rate of reduction in the matter contents from its bitons also will increase. Disturbances created by quanta of matter, released from the bitons, will be able to form photons of higher frequency. They will be radiated from the region. As the gravitational collapse of a large macro body continues, initially all radiations from its region will be of infrared frequencies. Gradually frequencies of radiations will include visible spectrum, ultra violet light, X-rays, cosmic rays, etc. This process does not require help from the assumed fusion of atoms. Hence all very large macro bodies emit radiations only due to their gravitational collapse.

Large macro bodies are formed from interstellar dust clouds present at different regions of space. Dust cloud in a very large region, along with small matter bodies in it, condenses and coagulates under apparent attraction due to gravitation. Depending on the matter distribution and homogeneity of the condensing dust cloud, it may condense to be a composite system of macro bodies of assorted sizes, called a galaxy. A single macro body, formed by very large concentration of 3D matter, within or outside a galaxy is a black hole.

If the total matter content in a region condenses to a black hole, 3D matter particles in it will be gravitationally squeezed to make the body's matter content very dense and increase its internal pressure to very high value. This may cause creation of very large number of photons, including very high frequency photons, in the region. By their inherent properties, these photons will be radiated from the region. However, if the matter content of the body is very high, apparent gravitational attraction between the radiated photons and the black hole will be very high.

Efficiency of an external effort on a matter body reduces as th linear speed of the body increases. External effort will be unable to act on a body moving at the linear speed of light. As a photon is moving at the speed of light, an external effort in the direction of its linear motion is unable to act on it. However, an external effort (or components of external efforts) in direction opposite to the direction of a photon' linear motion is fully active on a photon. Apparent gravitational attraction between a photon (which is moving away) and a black hole is fully effective on the photon. This external effort tends to slow down the photon. 2D energy fields about the photon sense photon's slowing and try to compensate it by a reduction in the matter content of the photon. Quanta of matter are released from the photon's core body. As per physical laws, a reduction in the mass of a body under external effort should counteract reduction in its linear speed. However, as the external effort (trying to slow down the photon) is continuous, action of slowing down the photon will continue until apparent gravitational attraction is reduced to negligible value (at very great distance between the black hole and the photon) or until whole of the photon's matter content is reverted to independent quanta of matter. Such radiations will not reach an outside observer. Photons of average and low frequencies from a black hole will not survive to reach an outside observer. This makes a black hole, an unobservable matter body. Hence, the name 'black hole'.

Reversion of 3D matter in the photons, radiated from a black hole, into free quanta of matter helps to recycle matter in the universe. Due to very high speed of photons, quanta of matter, freed from matter cores of photons, are distributed over a very large region in space. This prevents concentration of free quanta of matter to form disturbances and new photons. Instead, these free quanta of matter gradually migrate into surrounding 2D energy fields. 2D energy fields are inherently stable and serene entities, which represent a highly ordered system. Entropy in 2D energy fields can be approximated to nil value.

Development of 3D matter bodies increase entropy in the universe. Reversion of 3D matter into free quanta of matter and their eventual migration into highly ordered state of 2D energy fields helps to reverse entropy of universe. For a period, rate of development of 3D matter in the universe will dominate and total 3D matter content (and macro bodies) in the universe will gradually increase. Number of black holes in a region of space will reach a saturated state. They will revert 3D matter into the stable state of 2D energy fields at an increasing rate. For the following period, entropy of the universe will gradually reduce as more and more 3D matter is reverted to 2D energy fields. When the amount of 3D matter reduces to lower level, rate of development of 3D matter from 2D energy fields will increase. Cyclic conversion of matter into its 3D state and its reversion into 1D status in the universal medium keeps entropy of universe within limits. These cyclic changes maintain perpetual steady state of universe.

By the time photons of very high frequencies leave the influence of a black hole, their frequencies will be lowered considerably. Remnants from the high frequency photons, which lost their matter contents near black holes and remnants of photons, which lost their matter contents during their travel through extremely great distances, are observed by us as 'background radiations'.

* ** *** ** *

Index

Symbols
2D energy field 40

A
Absolute reference 42
Action at a distance 202
Aether drag 27
Apparent gravitational attraction 89
Attraction due to gravitation 99

B
Biton 147
Black hole 336

C
Central force 238

D
Dark matter 46
Deuteron 149
Distortion field 43, 144
Disturbance 66
   critical size 122
   ejection force 130
   internal pressure 119
   magnitude 69
   spinning force 132

E
Electric charge 146
Electric field 146
Electron 148
Energy 39

F
Field forces 54
Force 39, 47, 150

G
Galaxy 336
Gravitation 71
   repulsive nature of 173
Gravitational
   apparent attraction 173
   field 87, 150
   force 71
   pressure 71
Gravitational force 71

H
Halo 337
Hexton 148

I
Inertia 151
Inertial field 145

J
Jamming effect 87

L
Levitation 206

M
Magnetic field 145
Mass 221
Matter field 150

N
Nebular hypothesis 336
Neutron 148
Nuclear field 147

P
Photon 136
   axis of 137
   core body of 136
   critical disc size 135
   equator of 137
   inertial pocket 136
   poles of 137

Positron 149
Power 39
Proton 149
Push gravity 173

Q

Quantum of matter 30
  self-adhesion 31
  self-constriction 31
  self-elongation 31

R

Reactive force 47
Real matter
  creation of 123
Relativistic mass 98, 222
Rest mass 177, 221

S

Storm surges 318
String theories 33
syzygy 328

T

Tetron 148
Tide 315

U

Universal medium 65

W

Weight 225
Work 39, 150
Work-done 47

* ** *** ** *

Made in the USA
Middletown, DE
08 July 2017